AF615027

Contaminated Sites and Environmental Cleanup

CONTAMINATED SITES AND ENVIRONMENTAL CLEANUP

International Approaches to Prevention, Remediation, and Reuse

G. WILLIAM PAGE

Department of Planning
State University of New York at Buffalo
Buffalo, New York

ACADEMIC PRESS

San Diego • London • Boston • New York • Sydney • Tokyo • Toronto

Cover photograph courtesy of New York State Department of Environmental Conservation.

This book is printed on acid-free paper. ♾

Academic Press
525 B Street, Suite 1900, San Diego, California 92101-4495, USA
http://www.apnet.com

Academic Press Limited
24-28 Oval Road, London NW1 7DX, UK
http://www.hbuk.co.uk/ap/

Library of Congress Cataloging-in-Publication Data

Page, G. William (George William), date.
Contaminated sites and environmental cleanup : international approaches to prevention, remediation, and reuse / by G. William Page.
p. cm.
Includes index.
ISBN 0-12-543580-0 (alk. paper)
1. Hazardous waste sites--Case studies. 2. Hazardous waste site remediation--Case studies. I. Title.
TD1030.P34 1996
363.73'84--DC20 96-29220
CIP

PRINTED IN THE UNITED STATES OF AMERICA
96 97 98 99 00 01 EB 9 8 7 6 5 4 3 2 1

To my wife, Jo Beth Mertens, Ph.D.,
who provided invaluable advice,
support, and encouragement

Contents

CHAPTER 8

THE SUPERFUND PROGRAM IN THE UNITED STATES

CHAPTER 9

CONTAMINATION AND ENVIRONMENTAL CLEANUP IN WESTERN EUROPE

CHAPTER 10

THE NETHERLANDS CASE STUDY

CHAPTER 11

CONTAMINATED SITES AND ENVIRONMENTAL CLEANUP IN THE UNITED KINGDOM

CHAPTER 12

CONTAMINATION AND ENVIRONMENTAL CLEANUP IN CENTRAL AND EASTERN EUROPE

CHAPTER 13

CONCLUDING COMMENTS ON CONTAMINATION AND ENVIRONMENTAL CLEANUP ISSUES

ACKNOWLEDGMENTS

The contaminated sites and environmental cleanup problem has been a concern of mine for many years. Over the past twenty years, I have done research on different aspects of the problem, usually from a policy perspective. Because this problem is complex, there have been many people from different disciplines who have helped me understand the diverse aspects of this problem.

Some of the individuals deserve special mention. Dr. Michael Greenberg of Rutgers University introduced me to this problem and his continued work on related issues has been an inspiration to me ever since. Dr. Jo Beth Mertens provided invaluable advice on economic issues, policy perspectives, and the clarity of presentation throughout the book. Professor Robert Berger of the State University of New York at Buffalo provided helpful feedback on legal issues. Peter van Arnheim of the University of Nijmegen and Dr. Gert de Roo of the University of Groningen provided translated documents and information for the case study of The Netherlands. Professor Chris Wood of the University of Manchester and David Tye of Paul Butler Associates, Manchester, provided assistance with the United Kingdom case study. Dr. David Packer of Academic Press provided helpful editorial advice and patience while job changes and moves delayed the book.

INTRODUCTION

THE PROBLEM

Environmental cleanup is a term used to describe the process of remediating toxic substances contamination of soil and groundwater. There are many sources of hazardous and toxic contaminants: Mining, agriculture, commerical establishments, and households all release hazardous and toxic substances to the land. However, past and present industrial activities are the main cause of contaminated sites, these sites are the primary focus of this book.

The environmental cleanup of contaminated sites is a substantial problem throughout the world. Every country has contaminated sites, and in some of the most economically developed countries these sites number in the hundreds of thousands. Because of toxic contaminants, these sites are a threat to human health and the ecosystem even when only small concentrations of these substances are present. The problem is relatively recent and complex, and potential solutions are expensive. Furthermore, the environmental policy to address toxic contamination and environmental cleanup is highly contentious.

POLICY APPROACHES

This book discusses the policy approaches available to prevent toxic contamination, to contend with the environmental cleanup of contaminated sites, and to redevelop these remediated sites for productive uses. Such policy must address three issues: (1) stopping hazardous and toxic pollution from creating additional contaminated sites, (2) protecting humans and ecosystems from toxic substances, and (3) cleaning up contaminated sites so that they can be ecologically or economically productive.

In addressing these three fundamental issues, the book answers two questions concerning contaminated sites. How big a threat are contaminated sites to human health? What should we do with contaminated sites? There is broad agreement that actions are necessary to minimize exposure of human populations to toxic substances present on or under contaminated sites. Should we fence the property and take other measures that keep people off contaminated sites? If so, how do we ensure that existing contaminants do not migrate and contaminate additional sites? How do we ensure that these protections remain in force for generations into the future, when many contaminants will still be toxic? Should we attempt to remediate the contamination? If we decide to remediate, how clean must the site become, at what cost, and who should pay? Should we put the site to productive use after remediation is complete, and, if so, what uses are appropriate?

This book pays special attention to the issue of "environmental liability." In reference to contaminated sites, environmental liability refers to a determination of the party or parties responsible for causing the contamination and who is to be responsible for the costs of remediation. Many countries support the principle that the party that caused the contamination should pay for the cleanup. Unfortunately, in many instances it is impossible to prove who caused the contamination; multiple parties may be responsible, or the responsible parties may have insufficient resources to pay for the cleanup. There are many possible approaches to formulating a policy to determine environmental liability, and this issue is at the heart of the policy debate over environmental cleanup.

The term "reuse" refers to how countries will use contaminated sites in the future. Because the cost of remediation is so high, great numbers of contaminated sites have not been remediated and are not in productive use. Often, contaminated sites are located in urban areas with a well-developed physical and human infrastructure for transportation, power, and services, as well as a trained labor force. Those contaminated properties that are not in productive use are known as "brownfield sites." If environmental policy does not force the cleanup of contaminated sites, these properties not only remain unproductive, but often become a public health problem and attract unwanted social activities. This book describe the different policy approaches that countries are using to return remediated sites to productive use.

This book discusses environmental cleanup in its ecological, political, economic, and perceptual contexts. It reviews the main policy options that have been developed to contend with the problems related to contaminated sites. Experience in various countries demonstrates that all policies have both intended and unintended consequences. This book reviews legislation, experience, and successes or failures of environmental policy in the United States and in several European countries. This book also reviews the debate over the attempt to develop a common policy toward contaminated sites for the European Union.

MAIN THEMES

Five main themes run through this book: prevention, remediation, reuse, equity, and international competitiveness.

Prevention

The prevention of future contaminated sites is a critical issue, and most countries have legislation in place designed for this purpose. However, few countries currently have legislation to force the remediation of existing contaminated sites. Environmental policy affecting already contaminated sites is the central focus of this book. Such policy has a considerable bearing on future toxic chemical use and disposal practices, especially those that entail a risk of polluting. The type and extent of remediation of contaminated sites, who pays for remediation, the equity of assessing the cost of remediation, and the perceptions of this complex environmental problem all have powerful impacts on the effectiveness of policy to prevent additional sites from becoming contaminated. The book addresses the impact of the policy on potential future contamination in analyzing policy options to remediate already contaminated sites.

Remediation

Remediation refers to the process of environmental cleanup of contaminated sites and the techniques to decrease or eliminate contamination from soil, surface water, or groundwater. Although this book discusses remediation techniques, it focuses on the policy implications of remediation rather than on the technical methods. The complex policy interactions between environmental cleanup, prevention, reuse, equity, and international competitiveness are also considered. How clean must the property be after remediation is completed? This critical question has direct implications on the selection of the remediation technique, the cost of the clean-up, and the redevelopment potential of the property.

Reuse

Reuse refers to how countries will use contaminated sites after they have completed remediation or after they have made a conscious decision not to remediate. In virtually every case, some contamination will remain. In many cases, the goal of remediation may be only to contain the movement of contaminants on the site so that they do not spread to other land and expose additional people. Will countries restrict some or all future uses of sites to prevent exposure of people to contaminants? If so, what policies can governments use to protect future generations from exposure? What level of contamination will remain after remediation? Should the extent of remediation be contingent on the proposed future use of the site? Future residential use obviously requires "cleaner" land than many other land uses, but how can we control future land use and for how long?

Equity

Equity refers to "fairness." In this context, it refers to fair allocation of the costs of environmental cleanup. In many cases, we cannot identify the individual or firm responsible for causing the contamination, or we cannot legally prove responsibility, or the responsible party may be unable to pay for the remediation. Often, practices were legal at the time they caused contamination. Should the present owner of property, who had no part in causing the contamination, be responsible for the costs of remediation? The fairness of environmental policy in balancing these aspects of the contamination and cleanup problem and the perceptions of such fairness are critical issues. Since policies vary among countries, the issue of who pays for remediation and the equity of allocating the costs may affect international competitiveness.

The equity of policy also has a spatial component. In many economically advanced countries, great numbers of contaminated sites remain unproductive and abandoned, usually in older cities. These "orphan" or "derelict" properties influence the spatial pattern of urban development and redevelopment and may have a profound influence on the people living in the vicinity. The location of contaminated sites in minority neighborhoods has become an increasingly visible issue, which is now referred to as "environmental racism." Abandoned contaminated properties often become the loci of serious public health and social problems, while placing substantial financial burdens on local governments and influencing the location of new economic activity. Often, new activity locates in suburban political units on the periphery of metropolitan areas. The economic forces driving the globalization of the world economy and the industrial restructuring of mature economies interact with the problem of contamination and environmental cleanup. Together, these forces produce severe problems in some metropolitan areas and for certain segments of the population.

International Competitiveness

There is no international uniformity of policy toward toxic contamination and environmental cleanup. Some countries have no policy. Others are spending huge sums on remediation. Are expenditures for remediation of contaminated sites a wise investment from an international competitiveness perspective? These are concerns that policies of environmental cleanup in some countries are imposing costs on selected industries that make them unable to compete in the international marketplace. Environmental liability concerns may be distorting markets and directing investments to locations or activities that are not optimal.

The problem of environmental cleanup differs around the globe. The most economically developed countries of North America and Western Europe have thoroughly investigated this problem and have the most experience trying to solve it. These countries have relatively successful policies to prevent the creation of new contaminated sites. The countries of Central and Eastern Europe, on the other hand, often have worse problems of existing contaminated sites and do not yet have effective prevention or remediation policies. Many of the developing countries of Asia, South America, and Africa still do not have effective prevention or remediation policies. The globalization of the world's economy is shifting manufacturing and other heavy industry from the more developed to the developing countries. Therefore, it is critically important that developing countries implement effective prevention and environmental cleanup policies.

The countries of Central and Eastern Europe formerly had command–administrative economies, and they have great numbers of contaminated sites. These nations are attempting to resolve problems at the same time that they transfer ownership of enterprises and land from the government to the private sector, and some are developing unique and innovative policies concerning environmental cleanup. In many parts of the world where government control was prevalent, rapid private industrialization is increasing or about to accelerate. Effective policies to control toxic contamination and direct environmental cleanup will help these countries to minimize adverse environmental impacts.

ORGANIZATION OF THIS BOOK

Chapters 1–6 present aspects of the environmental cleanup problem. Chapter 1 discusses the characteristics of toxic contamination and reviews environmental cleanup and the Superfund program in the United States. Chapter 2 describes the remediation techniques used to clean up contaminated sites. Chapter 3 presents the public health aspects of the problem, and Chapter 4 addresses policy approaches to prevent the creation of additional contaminated sites. Chapter 5 considers the legal aspects of environmental cleanup, and Chapter 6 describes the economic, spatial, and social effects of contaminated sites and environmental cleanup.

Chapters 7–12 discuss toxic contamination and environmental cleanup problems and policies in specific countries or regions. Chapter 7 reviews environmental cleanup in developing countries, and Chapter 8 describes the United States' experience with the Superfund program. This was the first program for environmental cleanup in the world and it has influenced many other countries in their efforts to develop policies. Chapter 9 considers toxic contamination and environmental cleanup efforts in Western Europe, with special emphasis on the European Union. Chapter 10 is a case study of efforts in The Netherlands, which was the first country in Europe to develop an environmental cleanup policy. Chapter 11 is a case study of efforts in the United Kingdom, which has pursued a unique approach that emphasizes redevelopment of contaminated sites. In Chapter 12, toxic contamination and environmental cleanup in Central and Eastern European countries, including the unique problems resulting from their years with command–control economies, are discussed.

The final chapter reviews some of the important lessons that countries can learn by studying different approahces to the toxic contamination and environmental cleanup problem.

G. WILLIAM PAGE

CHAPTER 1

The Problem of Contamination and Environmental Cleanup

INTRODUCTION

This chapter outlines the general problem of contaminated sites and environmental cleanup and discusses the characteristics and sources of toxic substances contamination that cause the potential threats to both human and ecosystem health. A historical case study in the United States demonstrates how economic forces created this environmental problem, which has become an important political issue in many countries.

Defining the Problem

The term "contaminated sites" refers to parcels of land on which or under which hazardous and toxic substances exist under conditions that do not effectively confine their movement. The "toxic time bombs" or "toxic hot spots" that the media often refer to are the worst examples of contaminated sites, many of which are former hazardous waste disposal sites. Many thousands of other sites have less extensive toxic contamination, but they are also part of the contaminated sites and environmental cleanup problem. The toxic substances found at contaminate sites threaten human and ecosystem health in the present as well as in the future. The enormous cost of environmental cleanup contributes to the abandonment of many contaminated sites and the subsequent failure to return them to productive use. Cleaning up sites often costs far more than the value of the land.

There are large numbers of contaminated sites in all the economically developed countries and comparatively fewer in other countries. Most of these contaminated sites are the legacy of many decades of industrial development and military activities. The United states has approximately 600,000 sites contaminated with toxic substances, although many of these may not represent a serious human health threat (Office of Technology Assessment, 1983). Estimates for the number of suspected contaminated sites in The Netherlands range from 200,000 to 600,000 (Soczó *et al.*, 1992; Business Roundtable, 1993). In Germany, there are 250,000 suspected contaminated sites (Franzius, 1992). In the United Kingdom, there are 50,000–150,000 suspected contaminated sites comprising from 50,000 to 250,000 hectares (123,550 to 617,750 acres) of contaminated land (Denner, 1992a).

The major cause of such contamination is the production of large volumes of toxic substances in modern industrial and postindustrial societies. Most countries use toxic substances in many industrial processes, including those associated with defense industries, manufacturing, agriculture, and the production of commercial and household products. Table 1.1 presents data that are rough approximations of the hazardous waste

TABLE 1.1

INTERNATIONAL HAZARDOUS WASTE GENERATION[a]

Nation	Year of data	Amount (tons)	Amount per capita (tons)	GNP per capita ($)
Switzerland	1987	300,000	0.05	21,000
United States	1985	583,000,000	2.3	17,000
Japan	1983	1,540,000	0.01	16,000
West Germany	1988	7,150,000	0.1	15,000
Canada	1983	3,500,000	0.2	14,000
France	na	19,800,000	0.4	13,000
Austria	1984	2,700,000	0.4	13,000
Denmark	1985	154,000	0.03	11,000
Italy	1988	5,000,000	0.09	10,500
Netherlands	1987	580,000	0.04	9,000
United Kingdom	1988	5,500,000	0.1	8,000
Belgium	1988	1,650,000	0.2	8,000
Hungary	1986	2,000,000	0.2	7,500
Average			0.2	

[a]Modified from Schwab (1993).
These data are rough approximations only, because of data differences.

produced in several industrialized nations. The quality and comparability of these data vary. For instance, the value of 2.3 tons per capita for the United States is more than twice other estimates of 1 ton for each man, woman, and child (Barnett, 1994). Nonetheless, the table does provide interesting comparisons. Even for countries with relatively comparable gross national product (GNP) per capita, there is surprising variation in hazardous waste production.

In most cases, gradual releases of toxic substances cause contaminated sites. Many people think that catastrophic single events, such as a fire at a plant, a spill of chemicals, or an illegal discharge such as "midnight dumping," are the common causes of contaminated sites. These events certainly do result in contaminated land. However, the gradual release of toxins to the environment through a routine discharge or leak over a long period of time cause the vast majority of contaminated sites.

Some activities that cause contamination are outside the obvious activities in a modern manufacturing/consumer society. Mining operations produce large quantities of toxic wastes. About 95% of the total material

TABLE 1.2

TYPES OF SITE AND CONTAMINATION AT U.S. SUPERFUND SITES[a]

Type of site or contamination	Percentage of sites
Abestos	1
Battery recycling	1
Industrial landfill	11
Metals	6
Metals/organic chemicals	16
Mining waste	3
Municipal landfill	9
Munitions	1
Organic chemicals	15
Polychlorinated biphenyls (PCBs)	6
Pesticides	7
Metal plating	5
Radioactive waste	1
Solvents	11
Wood preserving	2
Multisource groundwater	5
Total	100

[a]Modified from U.S. General Accounting Office (1993, Table 2.1).

removed in mineral extraction is waste that often contains heavy metals. Defense facilities have created numerous contaminated sites through fuel spills and weapons production wastes. By 1991, the U.S. Department of Defense had identified 17,660 sites with potential toxic contamination of soil or groundwater (Rose, 1994). The U.S. Department of Energy is responsible for the environmental cleanup of perhaps 30% of the country's most serious contaminated sites, which became contaminated through nuclear weapons and research operations (Bredehoeft, 1993).

Some contaminated sites exist from long before modern industrial activity. Tanneries and printing operations dating from the Middle Ages produced contaminated sites that are still a problem today. Gas works operating 100 year ago in many European cities produced volatile aromatic hydrocarbon, polynuclear aromatic hydrocarbon, and cyanide contamination (see the Tilburg Gas Works Case Study in Chapter 10).

A review of data concerning the contaminated sites in the U.S. Superfund program provides insight into the types of sites and their constituent contaminants. Investigators examined 149 Superfund sites at which environmental cleanup was complete or nearly complete in 1993 and found a wide variety of contamination types and sources (Table 1.2).

CHARACTERISTICS OF TOXIC CONTAMINATION

Several definitions are necessary to understand the problem of contamination and environmental cleanup. A "toxic" substance is dangerous to human health, usually depending on the precise level of concentration, compounding, or ionization of the chemical. A "hazardous" substance includes those with any one of the following four characteristics: toxicity, corrosivity, flammability, or reactivity. The distinction between pollution and contamination is also useful. Both terms refer to unwanted, human-caused changes in the physical, chemical, or biological characteristics of the natural environment. For the purposes of this book, pollution refers to substances released to the environment in the present, whereas contamination refers to an existing condition resulting from pollution at a previous time. Thus someone spilling used motor oil on the ground is creating pollution. If no one contains or removes the spilled oil from the soil, it becomes contamination. Since several toxic substances are present in used motor oil, the land on which someone spills oil becomes "contaminated land."

Not all contamination is the same. Contamination with toxic substances has some properties that are different from those of other types of contamination. Unlike conventional pollutants, many toxic substances are extremely persistent in the environment and a threat to human and ecosystem health at low levels of exposure. These long-lived toxic chemicals may be heavy metals, radioactive substances, or organic chemicals with toxic properties. Natural processes can disperse and degrade conventional pollutants in relatively short periods of time. For example, the discharge of domestic sewage to water bodies or the release of conventional pollutants to the atmosphere is ameliorated by the activity of bacteria, dispersion, and other natural processes. In contrast, toxic substances contamination in the environment may remain a serious problem for long periods, even if no additional pollution occurs.

The persistence of toxic substance contamination is critical to understanding the contamination and environmental cleanup problem. Heavy metals such as lead and mercury are primary chemical elements that will never degrade. The persistence of radioactive substances varies greatly, but some highly toxic radionuclides are extremely persistent. For instance, plutonium-239 has a half-life of 24,100 years (Eisenbud, 1987, Appendix). Some organic chemical compounds are chemically stable and therefore resistant to environmental degradation. Several developed countries ban the use of some of the most persistent pesticides, such as the heavily chlorinated hydrocarbon DDT. Highly persistent toxic chemicals are a special threat to ecosystems and human health. From the present until far in the future, they may migrate through poorly understood environmental pathways and thereby expose additional ecosystems and humans. Some of these toxic substances tend to bioaccumulate in the food chain.

The health risk of some toxic substances changes when human activities release them on the land or into the ground. Sunlight and exposure to oxygen increase the rate of degradation of many organic chemicals. Once organic compounds are in the soil or groundwater, they may have protection from such exposure, which can increase their persistence and thus the potential for human health and ecosystem damage.

Though industrial activity, nuclear energy, and the military have caused the greatest amount of contaminated land in densely developed areas, other activities can also cause contaminated sites. Even activities that release only small quantities of toxic chemicals are a problem, for many chemicals are toxic at extremely low levels of exposure. Evidence suggests that some toxic substances cause cancer at exposures of a few parts per billion or less. This means that it is potentially possible for a few gallons of toxic waste to contaminate billions of gallons of groundwater. Such contamination can be high enough to produce cancer in humans who drink the water over a lifetime.

Small quantities of toxic waste are significant dangers and many small commercial establishments and households can cause extensive and dangerous contamination. Dry cleaners, chrome platers, and diverse machine shops, printers, and others that use solvents have the potential to seriously contaminate soil and groundwater. Household wastes, including used motor oil, paints and thinners, cleaners, lawn care products, and other goods, contain toxic substances that can cause contamination. Mining, agriculture, and other activities also result in extensive toxic contamination of land. This book concentrates on sites contaminated from industrial sources in urban areas that threaten exposure of large human populations. However, environmental policy to deal with contaminated sites must apply to all contaminated sites regardless of location or source of contamination.

Contaminated sites themselves are only the beginning of the problem of toxic contaminants released to the natural environment. Once free in the environment, toxic substances can follow environmental pathways and tend to accumulate in what scientists refer to as "environmental sinks." Wetlands and the sediments at the bottom of rivers, estuaries, and harbors are such environmental sinks for toxic substances. In the Great Lakes region of North America, the International Joint Commission representing the United States and Canada has identified 362 different toxic chemicals in river and harbor sediments. These locations are the overwhelming majority of the "toxic hot spots" designated for remedial action.

The policy debate must be considered in a historical framework, because much of the toxic pollution that created today's contaminated sites occurred in past decades. Economic factors, public awareness of environmental issues, knowledge of the health effects of microcontaminants, and environmental legislation have changed over time. For decades when we had less knowledge of environmental hazards, industry routinely used many toxic or hazardous substances in ways that we now know caused contamination. Because society did not recognize the threat of toxic substances contamination, standard operating procedures for the use and disposal of materials and waste products were lax compared to today's environmental regulations. In the past, we did not realize that policies implemented to prevent toxic pollution are much more effective at protecting the environment and public health than policies to promote environmental cleanup. The former are also much less expensive than policy implemented to remediate toxic substances contamination.

In many developing countries with less experience with toxic contamination, environmental legislation is still lax compared to that of the United States and other economically developed countries. As the developing countries assume a rapidly increasing proportion of the world's industrial activity, they risk creating large contamination and environmental cleanup problems. Some developing countries are implementing or considering innovative policies to control hazardous wastes and prevent contaminated sites (see the discussion of Thailand's approach in Chapter 7).

CONTAMINATION AND ENVIRONMENTAL CLEANUP EXPERIENCE IN THE UNITED STATES

The following case study of the U.S. experience provides historical context to the problem of contamination and environmental cleanup. The United States is a useful case study because it has been at the forefront of environmental science and environmental policy over the past several decades. It was the first to identify contaminated sites as a serious problem and to develop innovative policies. Many of the most economically developed countries have only recently formulated policies to address contamination and environmental cleanup (Table 1.3).

The U.S. Superfund program is at the center of the policy debates about remediating contaminated sites. There are presently about 2000 sites on the program's National Priority List for cleanup. Approximately four million Americans live within one mile of a Superfund site and more than 40 million Americans live within four miles (National Research Council, 1991, p. 2). Over the next 30 years, researchers estimate that the United States will spend $400 billion to $1.7 trillion in remediating contaminated sites (Russel *et al.*, 1991). Superfund is a controversial program that has not solved the problem of contaminated sites and environmental cleanup, and the program's costs far exceed expectations. Policies for remediating contaminated sites controlled by the Departments of Energy and Defense continue to evolve, but are less successful at remediation than the Superfund program.

TABLE 1.3

INITIATION OF POLICY APPROACHES TO CONTEND WITH CONTAMINATED SITES

Country	Year	Source
United States	1976	See Chapter 1
Netherlands	1980	Visser (1993, p. 45)
Germany	1981	Kingsbury and Bingham (1992, p. 248)
Finland	1981	Visser (1993, p. 67)
Denmark	1983	NATO/CCMS (1992, p. 4)
Switzerland	1983	Visser (1993, p. 73)
Sweden	1985	Kingsbury and Bingham (1992, p. 191)
Norway	1988	Folkestad (1992)
Austria	1989	Kasamas (1992)
Canada	1989	Hill (1992)
France	1989	Goubier (1992)
United Kingdom	1990	Denner (1992b)

Prior to the heightened awareness of environmental problems in the 1970s, the United States exerted little control over the use and disposal of toxic substances. For example, the U.S. Solid Waste Disposal Act of 1965 and the Resource Recovery Act of 1970 treated the disposal of toxic and hazardous materials no differently than other waste products. Industry sent most waste to land disposal sites that were not designed to keep toxic and hazardous materials from escaping into the environment. Some land disposal sites were commercial or municipal landfills, but many were private disposal sites operated by an industry on the same property as the production facility.

Toxic substances pollution was and remains a serious problem in the United States. The Environmental Protection Agency (U.S. EPA) estimated that in the 1970s industry sent about 240 million tons of solid industrial wastes to land disposal sites each year. As much as 15% of this waste was hazardous (U.S. Environmental Protection Agency, 1977a,b). Annual waste production in the 1970s included 1700 billion gallons of liquid wastes that were pumped to some form of surface impoundment (Russell, 1978). A U.S. EPA survey of surface impoundments at the time found that the large majority were unlined with any material that might keep waste products from percolating into groundwater (U.S. Environmental Protection Agency, 1978). The U.S. government estimated that before the 1970s industries improperly disposed of 90% of all hazardous wastes in open pits, surface impoundments, vacant land, farmlands, and water bodies (U.S. Environmental Protection Agency, 1974).

Inadvertent leaks during transport, storage, or use can also create contaminated sites. Until the mid-1970s, U.S. firms annually released additional huge quantities of toxic and hazardous materials to the ground from leaking underground storage tanks and from accidental spills and leaks. Even after years of increasingly stringent regulation, EPA estimated that there are several million underground storage tank systems that contain petroleum or other hazardous substances and that as many as 25% of them leak (Evans, 1988). It is important that local governments in all countries have well-developed emergency preparedness plans to respond to the inevitable accidental spills and leaks of hazardous materials (United Nations Environment Programme, 1992; Organization for Economic Cooperation and Development, 1991).

It is widely believed that the free market economic system in the United States maximizes economic benefits to the nation. This assumes that if each firm and individual attempts to maximize their own self-interest, then the aggregate of this behavior will produce the greatest benefit to the nation. Prior to the increase in environmental awareness of contaminated sites in the mid-1970s, private firms maximized their private benefit.

Under then existing laws, firms could use, store, and dispose of toxic substances in ways that minimized costs. Unfortunately, these activities caused substantial toxic pollution. The aggregate benefit of this behavior was not socially or economically optimal because individual decisions did not include consideration of the true costs of pollution. The cost of remediating contaminated sites today is many times greater than the cost would have been to have proper controls on toxic substances at the time they polluted the environment.

In the United States, the private market economic system imposed no disposal costs on firms that dumped their toxic and hazardous wastes on or into the ground of their own property. Very few data for on-site disposal activities are available. For this reason, we do not have an accurate inventory of contaminated sites. Some firms know that contamination exists on their property, but they do not want that knowledge to become public. Without publicly acknowledging the contamination, some firms leave large portions of their property undeveloped and often erect fences to keep their employees and the public off the property. Often, the only way the public or government agencies can learn of contaminated private land is if the firm places the property on the real estate market. Real estate transactions for industrial land today require environmental assessments to discover if toxic contamination is present.

CONTAMINATION AND ENVIRONMENTAL CLEANUP BECOME A POLITICAL ISSUE

As mentioned earlier, in the 1970s the public awareness of toxic contamination grew in the United States. The residential neighborhood known as Love Canal in Niagara Falls, New York, was the single most notorious case to heighten such awareness. Before the Love Canal episode, the general public was not aware of the explosive growth in the production of new chemical compounds since World War II. Nor did the public realize that people could be unknowingly exposed to these chemicals because they were so poorly controlled, or that many of these chemicals were a threat to human health at levels of exposure that they could not see, smell, or taste. Love Canal dramatically changed the public's perceptions. Analogous cases of contaminated sites are the village of Lekkerkerk (see Chapter 10) and the Merwedepolder housing project in Dordrecht (Kingsbury and Bingham, 1992), both in The Netherlands.

From 1942 to 1953, the Hooker Chemical and Plastics Corporation had buried 22,000 tons of chemical waste in canals on its property. In the early 1950s, the company filled the canals, subdivided the property, and began selling the contaminated land. Developers bought the property and built a residential neighborhood known as Love Canal on this former chemical waste disposal site. By the mid-1970s, the media started reporting complaints by the residents of the Love Canal neighborhood about strange health problems that they attributed to the contaminated land (Leonard *et al.*, 1977). Health studies reported disturbing findings of cell aberrations and implications of increased risks of cancer, birth defects, and spontaneous abortion (New York Department of Public Health, 1978; Picciano, 1980; Deegan, 1987).

In August 1978, government officials declared the area unsafe. They evacuated 1004 households from the Love Canal area and spent $30 million purchasing homes in the most contaminated areas. Extensive publicity about Love Canal dramatically raised the public's perceptions of the problem of contamination and environmental cleanup.

Resolution of the legal actions concerning over Love Canal continued for many years. In June 1994, Occidental Chemical Company, the successor corporation to Hooker Chemical, agreed to pay $98 million to the state of New York for the state's part in the cleanup. This payment ended a 14-year-long lawsuit. In 1995, Occidental agreed to pay the U.S. EPA $129 million as settlement for EPA expenses at the Love Canal site.

Another example of a contaminated site that gained wide notoriety was the Valley of the Drums in Bullitt County, Kentucky. In 1975, investigation of the uncontrolled industrial waste dump in a rural, 13-acre valley found over 17,000 drums, many of them filled with hazardous waste. Deteriorated and leaking drums had released 140 different chemical compounds, including heavy metals, polynuclear aromatic hydrocarbons, and polychlorinated biphenyls (U.S. Environmental Protection Agency, 1992), and contamination was entering Wilson Creek, a tributary of the Ohio River. The devastation to the natural vegetation of the site made a powerful image of the ecological cost of toxic contamination.

Media coverage of such cases produced rapid and intense public concern. Researchers who investigated the flagship news programs of ABC, CBS, and NBC found that, between 1978 and 1987, the networks ran 99 Love Canal news stories that used 191.2 minutes of prime time television (Greenberg and Wartenberg, 1990). In addition to the cases of contamination and environmental cleanup that made national news, most regions of the United States also had media coverage of local contamination problems. For example, in Buffalo, New York, citizens discovered that the city had built a neighborhood playground on arsenic-contaminated soil (Thigpen, 1993). In all cases, the public demanded action from public officials.

CONCLUSIONS

The contamination and environmental cleanup problem requires three policy approaches: (1) take immediate action to protect the health of people at risk because of proximity to contaminated sites; (2) stop additional pollution from creating additional contaminated sites; and (3) clean up contaminated sites to render them environmentally healthy and available for productive uses.

Several countries have recognized the importance of this problem and have developed policies that attempt to solve it. All of these policy approaches have had some success in protecting people and ecosystems, stopping additional toxic pollution, and implementing environmental cleanups. However, no country has yet developed policies that completely eliminate the problem; contamination is too extensive and cleanup is too expensive.

The next three chapters cover the aforementioned policy approaches. Chapter 2 discusses the techniques used to achieve environmental remediation; Chapter 3 addresses exposure, toxicity, and risk issues regarding proximity to contaminated sites; and Chapter 4 discusses how to prevent additional contaminated sites.

References

Barnett, H. C. 1994. *Toxic Debts and the Superfund Dilemma.* Chapel Hill, NC: University of North Carolina Press.

Bredehoeft, J. D. 1993. Waste remediation: A 21st-century issue. *Forum Appl. Res. Public Policy* **8**(1), 135–139.

Business Roundtable. 1993. *Comparison of Superfund with Programs in Other Countries.* Washington, DC: Business Roundtable, 117 pp.

Deegan, J. Jr. 1987. Looking back at Love Canal. *Environ. Sci. Technol.* **21**(5), 421–426.

Denner, J. 1992a. Contribution to the Tour-de-Table: United Kingdom. *Summary Report.* The 1992 NATO/CCMS Pilot Study Meeting on Evaluation of Demonstrated and Emerging Technologies for the Treatment and Clean-up of Contaminated Land and Groundwater, Budapest, October.

Denner, J. 1992b. UK Tour-de-Table paper. *Summary Report.* The 1992 NATO/CCMS Pilot Study Meeting on Evaluation of Demonstrated and Emerging Technologies for the Treatment and Clean-up of Contaminated Land and Groundwater, Budapest, October.

Eisenbud, M. 1987. *Environmental Radioactivity: From Natural, Industrial, and Military Sources,* third edition. New York: Academic Press, 475 pp.

Evans, J. 1988. *Musts for USTs.* Washington, DC: U.S. Environmental Protection Agency.

Folkestad, B. 1992. Tour-de-Table—Norway. *Summary Report.* The 1992 NATO/CCMS Pilot Study Meeting on Evaluation of Demonstrated and Emerging Technologies for the Treatment and Clean-up of Contaminated Land and Groundwater, Budapest, October.

Franzius, V. 1992. Recent developments in national programs, Federal Republic of Germany. Contribution to the Tour-de-Table. *Summary Report.* The 1992 NATO/CCMS Pilot Study Meeting on Evaluation of Demonstrated and Emerging Technologies for the Treatment and Clean-up of Contaminated Land and Groundwater, Budapest, October.

Goubier, R. 1992. Tour-de-Table: French Presentation. *Summary Report.* The 1992 NATO/CCMS Pilot Study Meeting on Evaluation of Demonstrated and Emerging Technologies for the Treatment and Clean-up of Contaminated Land and Groundwater, Budapest, October.

Greenberg, M., and D. Wartenberg. 1990. How epidemiologists can improve television network news coverage of disease clusters reports. *Epidemiology* **1,** 168–170.

Hill, G. H. 1992. An overview of Canada's National Contaminated Sites Remediation Program. *Summary Report.* The 1992 NATO/CCMS Pilot Study Meeting on Evaluation of Demonstrated and Emerging Technologies for the Treatment and Clean-up of Contaminated Land and Groundwater, Budapest, October.

Kasamas, H. 1992. Contribution to the Tour-de-Table: Austria. *Summary Report.* The 1992 NATO/CCMS Pilot Study Meeting on Evaluation of Demonstrated and Emerging Technologies for the Treatment and Clean-up of Contaminated Land and Groundwater, Budapest, October.

Kingsbury, G., and T. Bingham. 1992. *Reclamation and Redevelopment of Contaminated Land: Vol. II: European Case Studies* (EPA 600/R-92-031). Cincinnati, OH: U.S. Environmental Protection Agency, Office of Research and Development, Risk Reduction Engineering Laboratory.

Leonard, R. P., P. H. Wetham, and R. C. Zeigler. 1977. *Characterization and Abatement of Groundwater Pollution from Love Canal Chemical Landfill, Niagara Falls, New York.* Buffalo, NY: CALSPAN Corporation.

National Research Council. 1991. *Environmental Epidemiology: Vol. 1: Public Health and Hazardous Waste.* Washington, DC: National Academy Press.

NATO/CCMS. 1992. *Summary Report.* The 1992 NATO/CCMS Pilot Study Meeting on Evaluation of Demonstrated and Emerging Technologies for the Treatment and Clean-up of Contaminated Land and Groundwater, Budapest, October.

New York Department of Public Health. 1978. *Love Canal: Public Health Time Bomb, Special Report to the Governor and Legislature.* Albany.

Office of Technology Assessment. 1983. *Technologies and Management Strategies for Hazardous Waste Control.* Washington, DC: Congress of the United States, March.

Organization for Economic Cooperation and Development (OECD). 1991. *Draft Guiding Principles for Chemical Accident Prevention, Preparedness and Response.* Paris: Organization for Economic Cooperation and Development, Environment Directorate, Environment Committee.

Picciano, D. 1980. *Pilot Cytogenic Study of Love Canal, New York.* Washington, DC: U.S. Environmental Protection Agency, prepared by the Biogenics Corporation.

Rose, D. 1994. DOD, DOE clean-up efforts promising. *Environ. Prot.* **5**(1), 58–64.

Russel, M., E. Colglazier, and M. English. 1991. *Hazardous Waste Remediation: The Task Ahead.* Knoxville, TN: Waste Management Research and Education Institute.

Russell, C. (ed.). 1978. *Safe Drinking Water: Current and Future Problems.* Washington, DC: Resources for the Future.

Schwab, J. 1993. Toxics use reduction. *Environment & Development.* American Planning Association, March.

Soczó, E., T. A. Meeder, and C. W. Versluijs. 1992. Ten years of soil clean-up in the Netherlands. Contribution to the Tour-de-Table. *Summary Report.* The 1992 NATO/CCMS Pilot Study Meeting on Evaluation of Demonstrated and Emerging Technologies for the Treatment and Clean-up of Contaminated Land and Groundwater, Budapest, October.

Thigpen, D. 1993. The playground that became a battleground. *Nat.. Wildl.* February–March.

United Nations Environment Programme. 1992. *Hazard Identification and Evaluation in a Local Community.* Industry and Environment/Programme Activity Center (IE/PAC), Technical Report No. 12. Paris: UNEP.

U.S. Environmental Protection Agency. 1974. *State Decision-Makers Guide for Hazardous Waste Management* (SW-612). Washington, DC: U.S. Government Printing Office.

U.S. Environmental Protection Agency. 1977a. *Waste Disposal Practices and Their Effects on Ground Water (Report to Congress).* Washington, DC: U.S. Government Printing Office.

U.S. Environmental Protection Agency. 1977b. *State Decision-Makers Guide for Hazardous Management.* Washington, DC: U.S. Government Printing Office.

U.S. Environmental Protection Agency. 1978. *Surface Impoundments and Their Effects on Ground-Water Quality in the United States—A Preliminary Survey* (EPA 570/9-78-044). Washington, DC: U.S. Government Printing Office.

U.S. Environmental Protection Agency. 1992. *Superfund at Work: Valley of the Drums Cleanup: A Superfund Benchmark* (EPA 520/F-92-006). Washington, DC: Office of Solid Waste and Emergency Response.

U.S. General Accounting Agency. 1993. Superfund Cleanups Nearing Completion Indicate Future Challenges. GAO/RCED-93-188, Washington, DC: Government Printing Office.

Visser, W. J. F. 1993. *Contaminated Land Policies in Some Industrialized Countries.* Report to the Technical Soil Protection Committee, The Hague.

CHAPTER 2

REMEDIATION

INTRODUCTION

Remediation refers to the actions taken to clean up contamination and return the site to its original condition before the pollution event. The term derives from the word remedial and implies actions that are restorative, healthful, or salutary. A variety of remediation techniques are available depending on the quantity and characteristics of the specific contaminants present and on the geology and geochemistry of the site.

Environmental policy in several countries requires remediation of some contaminated sites, and experience in the past decade has produced considerable improvement in existing remediation techniques. New techniques to discover and identify the extent and type of contamination may become valuable tools. For example, ground-penetrating radar shows promise in identifying and therefore contributing, in a cost-effective way, to remediation of subsurface contamination. However, remediation is often inefficient, not cost-effective, and not available for certain contaminants.

The goals of remediation for specific contaminated sites may vary in terms of the degree of cleanup sought. Some remedial actions may seek only to contain contaminants on the site. At the other extreme, remedial actions may attempt to clean up the contaminated property with the goal of restoring it to its precontamination condition. Permanent solutions are preferable to containment from environmental and economic development perspectives, but they are often much more expensive. Long-term monitoring of sites may be necessary to ensure that the remediation achieved the cleanup goals. Although 100% restoration is impossible, remedial actions do clean up some contaminated properties to a degree that permits the safe use of the site for any activity.

The cost of remediation is high and, unfortunately, the cost is not linear. In the United States, the average Superfund site remediation costs

$26 million (U.S. General Accounting Office, 1991). As remediation projects approach higher percentages of contaminant cleanup, the marginal cost of remediation increases dramatically. The cost to increase the percentage of contaminant eliminated from 90 to 95% may be many times more than the cost to remove the first 90% of contamination. Thus remediation costs are a critical issue in the policy question, "How clean must a site be after remediation?"

REMEDIAL ACTIONS INVOLVING CONTAINMENT

Containment is one of the most common remediation techniques, whose goal is to confine the contaminants present in order to limit their contact with people, the ecosystem on the site, and their migration off the site. While contaminants are unconfined, they may migrate from the original contaminated site and expose humans and the ecosystem over a broader and often poorly predited area. Containment, treatment on-site, or even temporary storage of toxic substances on-site prior to removal often incite strong opposition. People are fearful that the environmental cleanup treatment may inadvertently expose them to health risks (see the discussion of risk perception in Chapter 3).

Remediation projects employ different containment techniques. Projects can excavate and remove contaminated soil to a secure landfill. To meet hazardous waste standards, such landfills must have impermeable synthetic liners and monitoring systems designed to confine contaminants. However, over a long time horizon, containment is not a final remediation because even the most elaborate and secure landfill will eventually leak. In the United States, monitoring of toxic chemicals in groundwater found releases at 74% of the hazardous waste land disposal facilities (U.S. General Accounting Office, 1995).

Environmental cleanup using containment can involve three variations: (1) containment in-place, (2) excavation and movement to contain in a different location on the site, and (3) excavation and movement to contain the contaminated soil in a different location off-site. An example of containment in-place is the Powersville Superfund site in Peach County, Georgia. Over the years, people disposed of hazardous wastes in an old quarry used as a municipal landfill (U.S. Environmental Protection Agency, 1993a). The wastes included pesticides, vinyl chloride, and several heavy metals. The pesticides from the dump site entered groundwater and threatened people, orchards, crops, and livestock (Fig. 2.1). In 1987, environmental cleanup included installing a multilayer synthetic cap to cover the 15-acre site, which prevents rainwater from mobilizing contaminants in the

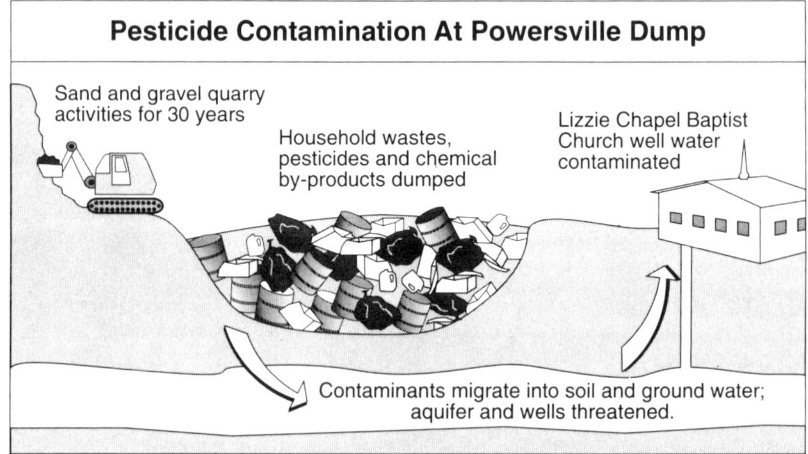

FIGURE 2.1 Pesticide contamination at Powersville Dump, Peach County, Georgia. [From U.S. Environmental Protection Agency (1993a).]

landfill. Subsequent extension of municipal water lines to area residents further limits their exposure to the contaminated groundwater.

An example of excavation and movement of contaminated materials and soils to a different location on the site is Pesses Company S'West Superfund site in Fort Worth, Texas. This 4.2-acre site contains a metals reclamation facility, which the owners abandoned in 1981 (U.S. Environmental Protection Agency, 1995). The site contained heavy metals, including cadmium, copper, lead, and nickel. The potential health risk was high because 19,500 people lived within one mile of the site. In 1992, crews excavated 12,000 cubic yards of contaminated soil, which they moved to a different location on the site for treatment. Treatment involved mixing the contaminated soil with cement kiln dust to create a solidified mixture, and then covering the solidified mixture with a high-density polyethylene liner and a steel-reinforced concrete cap.

Excavation and movement of contaminated materials and soils to a different location off-site is another remediation technique. Two examples are the Radium Chemical site in New York City and the Harvey and Knott Drum site in New Castle County, Delaware. The Superfund program calls these remedial actions "source removal." The Radium Chemical site covers only one-third acre in a light industrial area of the city and stored radium used in cancer treatment. The company used the site from 1951 until they abandoned the site in 1988, when New York State ordered the company to cease operations (U.S. Environmental Protection Agency, 1993b). The radium on the site represented a threat to the community of 300,000 who lived within three miles of the site. Environmental cleanup actions included removing 10,000 small metal containers with 120 curies of radium, decontamination of the building followed by its demolition, excavation of contaminated soil, and the removal of the contaminated sewer system (Fig. 2.2). The cleanup crew transferred the contaminated materials to an approved waste disposal site.

A private company used the 2-acre Harvey and Knott Drum site as an open dump and burning site for sanitary, municipal, and industrial wastes from 1963 to 1969 (U.S. Environmental Protection Agency, 1993c). In 1984, EPA removed 200 drums containing 1925 gallons of wastes as well as 500 empty drums form the site. The responsible party transported the drums and 955 cubic yards of soil to an off-site, licensed disposal facility for containment (Fig. 2.3). In addition, EPA capped selected "hot spots" of contaminated soil and began monitoring the groundwater.

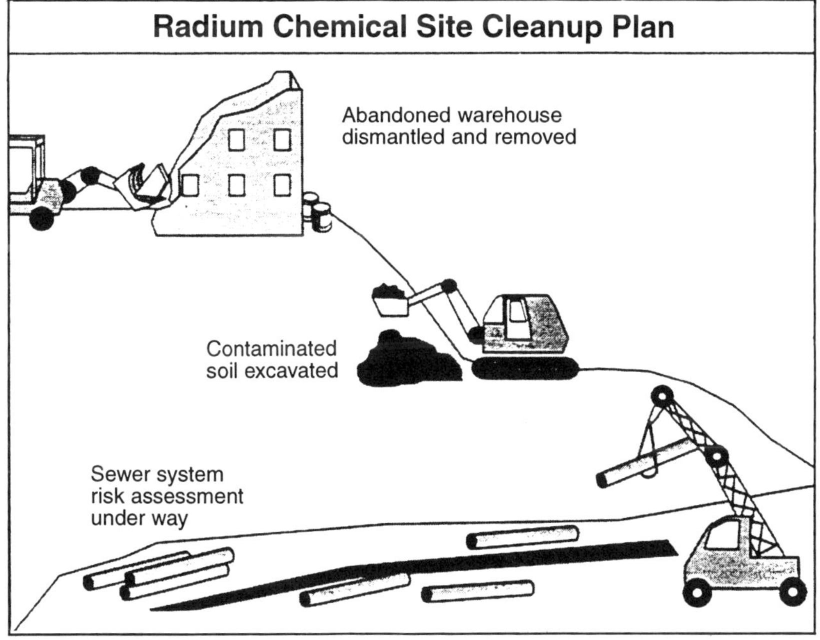

FIGURE 2.2 Radium Chemical site cleanup plan, New York City, New York. [From U.S. Dept. of Commerce, National Technical Information Service, Springfield, VA 22161.]

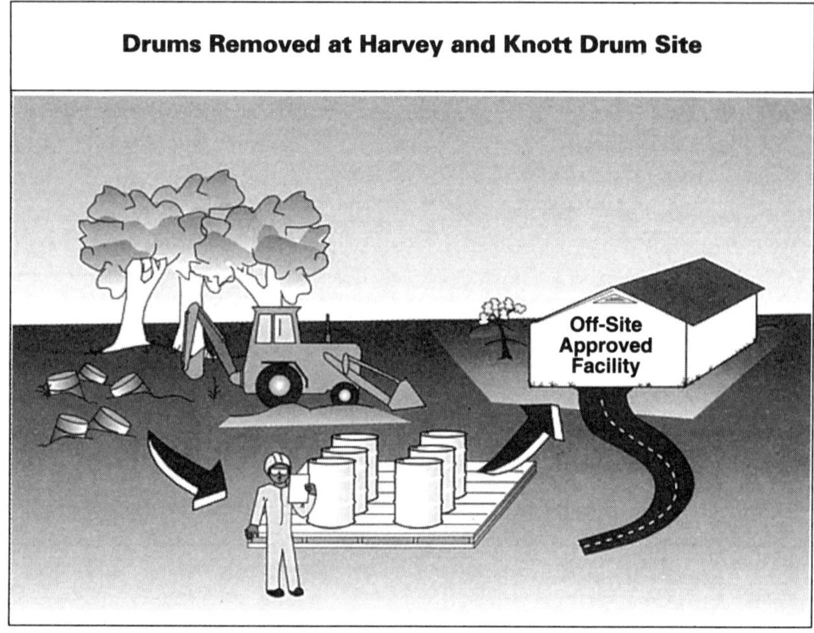

FIGURE 2.3 Harvey and Knott Drum site, New Castle County, Delaware. [From U.S. Environmental Protection Agency (1993c).]

Remediation projects can employ different containment techniques to confine contaminated soil or groundwater under the surface of the land. Physical structures, such as slurry walls, can confine contaminated groundwater when combined with pumping schemes. Pumping contaminated groundwater out of the aquifer can help confine the contamination on-site (McKay and Cherry, 1989). Furthermore, strategically pumping clean water into an aquifer can create groundwater flows that help confine the movement of a plume of contaminated groundwater (Cohen *et al.*, 1987). Other less common techniques are also available.

REMEDIAL ACTIONS INVOLVING TREATMENT

At the other extreme of remediation are techniques that attempt to remove or destroy the contamination. These techniques try to restore the land to its condition before human action released the toxic substances on the site (Rose, 1994; Kinner *et al.*, 1993; U.S. Environmental Protection Agency, 1990; Clark, 1987; Page 1987a,b). Environmental cleanup teams must be careful that the attempted remediation does not merely transfer contaminants from one medium to another. This could occur, for example, if soil contaminants leach into water or if incomplete incineration converts soil contaminants into toxic air pollutants. We will briefly review some of these treatment techniques, broadly categorized as thermal, thermal desorption, extraction, and bioremediation.

Thermal Treatments

Thermal treatment is one of the most widely used remediation techniques and usually involves incinerating contaminated soil to destroy organic compounds (U.S. Environmental Protection Agency, 1993d). It works well for many compounds, but some nations do not allow its use if chlorinated compounds are present because of potential combustion byproducts such as dioxins and furans.

An example of incineration to remediate contaminated soil is the environmental cleanup of the Celanese Corporation Superfund site in Cleveland County, North Carolina. This 469-acre site is an active polyester production facility that produces polyester chips, typewriter key caps, automotive components, and filament threads, which are used in apparel, home furnishing, and industrial products (U.S. Environmental Protection Agency, 1992a). Improper disposal, burning, and burial of wastes and sludge from 1960 to 1969 produced contaminated groundwater, soil, and sediment. The contaminants included volatile organic compounds (VOCs), such as ethylene glycol, phthalates, benzene, and trichloroethane, and heavy metals, among them chromium, arsenic, and lead. One phase of the environmental cleanup involved excavating 1800 cubic yards of sludge and incinerating it on-site in a mobile rotary kiln incinerator (Fig. 2.4). Incineration detoxifies the contaminated sludge by destroying organic compounds, often to the 99.99% removal level. Incineration has the added benefits of reducing the volume of wastes and converting the sludge to solids by vaporizing water and other liquids. Hazardous waste incinerators must maintain high temperatures in the range of 1800° to 2500°F.

Another example of using a mobile incinerator to remediate VOCs in bulk wastes and contaminated soil is the Big D Campground Superfund site in Ashtabula County, Ohio. From 1964 to 1976, responsible parties disposed of 14,000 drums of manufacturing wastes, including the VOC toluene, in an old sand and gravel quarry (U.S. Environmental Protection Agency, 1994a). The incinerator employs two combustion chambers, the first to separate the contaminants from the soil and the second to destroy the VOCs. The incinerator operates 24 hours per day, 7 days a week, and averages treatment of 235 tons per day.

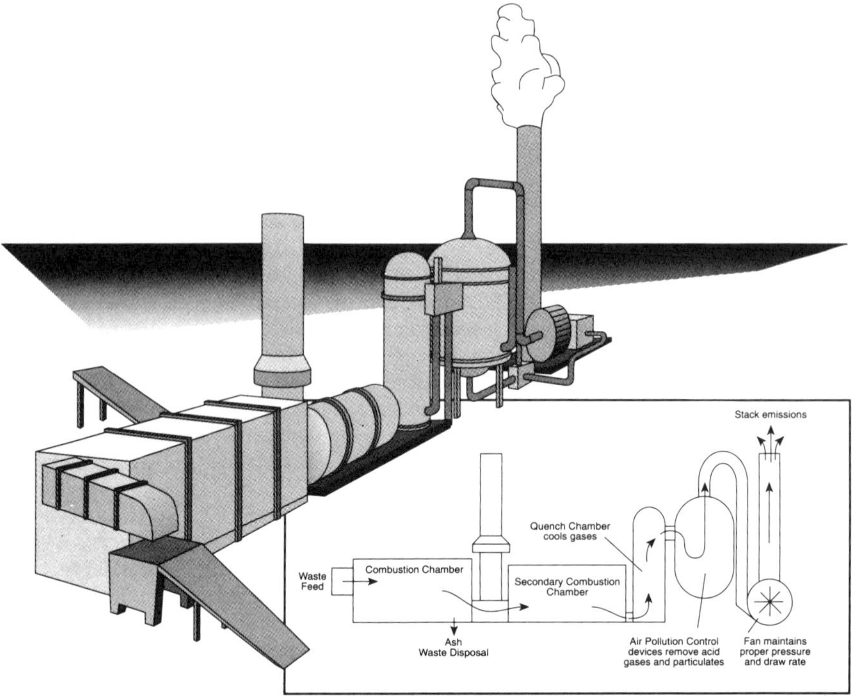

FIGURE 2.4 Rotary kiln incinerator. [From U.S. Environmental Protection Agency (1992a).]

A mobile incinerator at the Old Midland Products site in Yell County, Arkansas, achieved environmental cleanup rapidly. It averaged almost 17 tons of lagoon sludges and soil per hour and 85,000 tons in a 10-month period (U.S. Environmental Protection Agency, 1993e).

Some newer variations of thermal treatment, such as low-temperature thermal desorption, vaporize volatile and semivolatile organic compounds from soil, sludges, and other solids using low levels of heat that do not burn the contaminants. By avoiding incineration, these thermal treatments do not produce potentially dangerous combustion by-products.

Thermal Desorption Treatments

A variety of thermal desorption treatments are used to remediate soil contaminated with VOCs. These techniques use heat to remove the VOCs from the soil and then use condensation to capture the VOCs for disposal.

Such treatments do not attempt to burn the VOCs as incineration would. Thermal desorption is most widely used when polychlorinated biphenyls (PCBs) are present, because the process does not produce dangerous combustion by-products. Nonetheless, the EPA has not yet identified any technologies that it believes to be as effective as incineration for most PCB- or dioxin-contaminated sites (U.S. General Accounting Office, 1996a).

The American Thermostat Superfund site in South Cairo, New York, used thermal desorption among other remediation techniques. From 1954 until 1980, the company improperly disposed hazardous wastes on an 8-acre site, which resulted in soil contaminated with a variety of VOCs and other organic compounds, as well as arsenic, chromium, and lead (U.S. Environmental Protection Agency, 1992b). Catskill Creek, a trout stream, is less than one-quarter mile from the site and 5000 people live within three miles. Using thermal desorption, the remediation team heated the soil to about 400°C to vaporize moisture and VOCs, and a vapor extraction system then removed the VOCs from the gas (Fig. 2.5).

The CEC Superfund site in Bridgewater, Massachusetts, is another

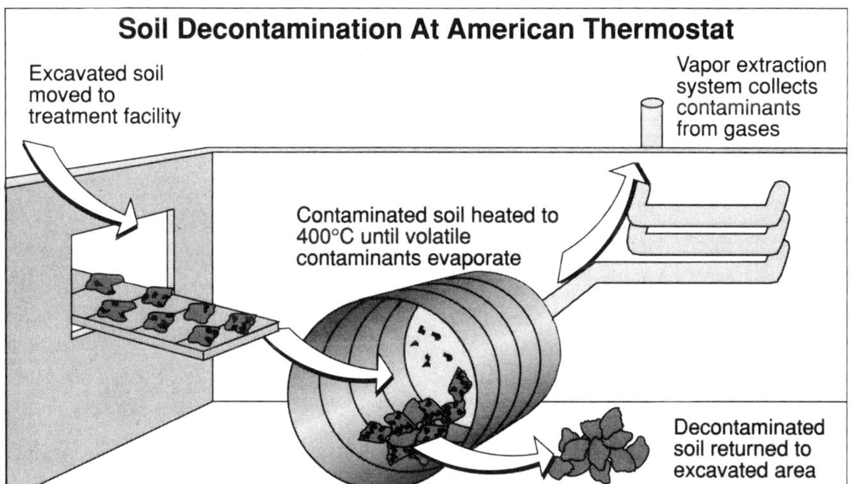

FIGURE 2.5 As knowledge about hazardous wastes has grown, the EPA has developed several methods for soil decontamination. Based on the type of contamination at the American Thermostat site, South Cairo, New York, the soil was treated with a low-temperature heat to evaporate the volatile organic compounds. [From U.S. Environmental Protection Agency (1992c).]

example of a site that used thermal desorption treatments to remove VOCs that did not contain PCBs. CEC operated a storage tank and an incinerator on a 6-acre site in an industrial park, where it stored waste oils, solvents, pesticides, industrial emulsions, lacquers, and electroplating wastes containing cyanide products (U.S. Environmental Protection Agency, 1994b). In this case, the soil was heated to temperatures between 300° and 500°F to vaporize the contaminants and the VOCs were then captured by passing the vapor through carbon adsorption beds. The treatment captured 1242 pounds of VOCs during the treatment of 11,330 tons of contaminated soil.

Treatment at the Re-Solve Inc. Superfund site in North Dartmouth, Massachusetts, used a low-temperature thermal desorption process to remediate soil contaminated with a variety of VOCs, including PCBs (U.S. Environmental Protection Agency, 1994c). The company processed solvents, waste oils, organic and inorganic liquids and solids, alkalies, and PCBs on the 6-acre site from 1956 to 1980. The company had previously dumped distillation residue, liquid sludge waste, and impure solvents into four unlined lagoons that drained into groundwater. The company had also spread waste oil on the site to control dust and land-farmed waste oils on a portion of the site. The Copicut River, which at this point is a designated state Wildlife Protection Area, is less than 500 feet from the site. The EPA listed the site on the National Priority List in 1983 and completed the Removal Action in 1987 (see Chapter 8 for a discussion of the Superfund program process). In 1993, remediation teams began low-temperature thermal desorption treatment of contaminated soil, which involved two steps: (1) the low-temperature thermal desorption process heats the soil to vaporize the PCBs and VOCs, collects the vapors, and condenses them into a concentrated liquid; and (2) dechlorination of the chemicals in the liquid reduces its toxicity prior to transport of the liquids to a hazardous waste landfill (Fig. 2.6).

Another Superfund site that utilized low-temperature thermal desorption treatment of contaminated soil is the Fulton Terminals site in Oswego County, New York. The company used the 1.6-acre site bordering the Oswego River as a tank farm to store millions of gallons of waste oils and sludges. Leaks and spills contaminated the soil, groundwater, and the river (U.S. Environmental Protection Agency, 1993f), primarily with VOCs and PCBs, although heavy metals were also present. About 13,000 people lived within three miles of the site, and contamination threatened people swimming in the river or eating fish from the river. Because of the presence of PCBs, the remediation team selected this technique to avoid creation additional combustion by-products (Fig. 2.7).

The Wide Beach Superfund site in Brant, New York, used a different thermal treatment technique to remediate PCB contamination. This site

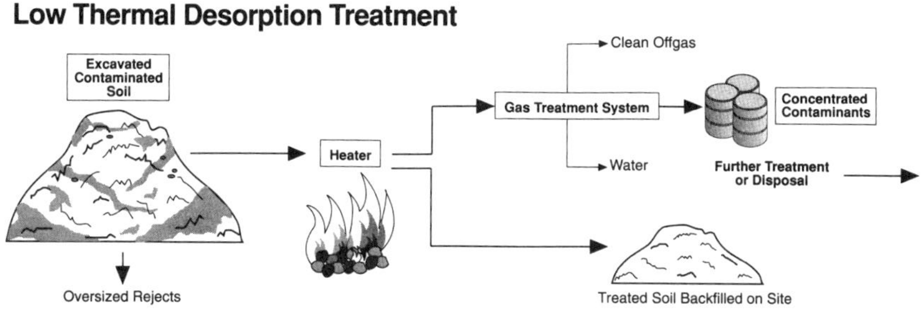

FIGURE 2.6 Low-thermal desorption treatment. [From U.S. Environmental Protection Agency (1994b).]

occupies 55 acres in a residential and resort area on Lake Erie. From 1968 to 1978, the community spread PCB-contaminated waste oil on the dirt roads in the development for dust control (Fig. 2.8) (U.S. Environmental Protection Agency, 1992c). The treatment is called the anaerobic thermal process, which uses heat to vaporize organic contaminants from mixtures of solids and water such as the mixture of soil, sludge, and sediments found at Wide Beach. The initial stage evaporates moisture and light oily materials, and the second stage treats the remaining solids in an air-free vessel with temperatures of 950° to 1150°F. These temperatures vaporize the PCBs and remaining VOCs, which are then removed and condensed. Some

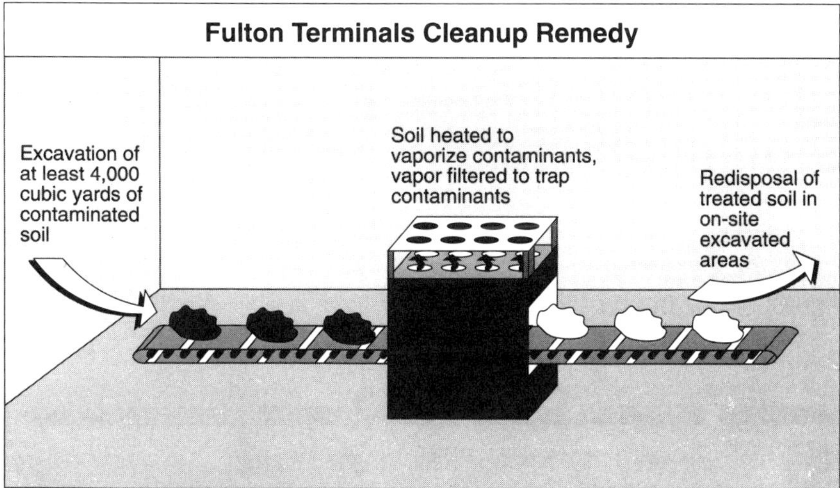

FIGURE 2.7 Fulton Terminals cleanup remedy, Oswego County, New York. [From U.S. Environmental Protection Agency (1993f).]

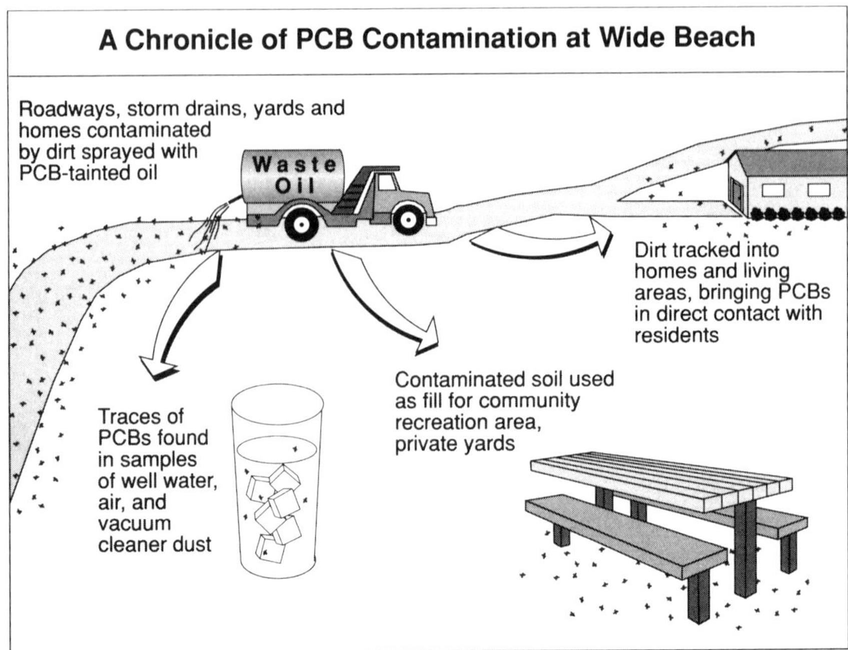

FIGURE 2.8 A chronicle of PCB contamination at Wide Beach, Brant, New York. [From U.S. Environmental Protection Agency (1992c).]

of the condensed organic chemicals can be recycled and others, such as PCBs, are sent to a hazardous waste facility.

Extraction Treatments

Extraction techniques physically remove the contaminated soil, surface water, or groundwater from a site and then confine or treat it. In some cases, remediation projects may extract and treat the contaminated soil, surface water, or ground water on-site. In other cases, remediation teams take the contaminated media somewhere else for disposal or treatment.

An example of remediation using extraction of surface water for treatment is the Valley of the Drums Superfund site in Bullitt County, Kentucky. This site is one of the most infamous examples of a contaminated site in the United States (see Chapters 1 and 8 for additional discussion of this site). One of the techniques that remediation teams used at the site involved collecting and treating contaminated surface water (Fig. 2.9), which was held in a lagoon until they could treat it. The goal was to keep the

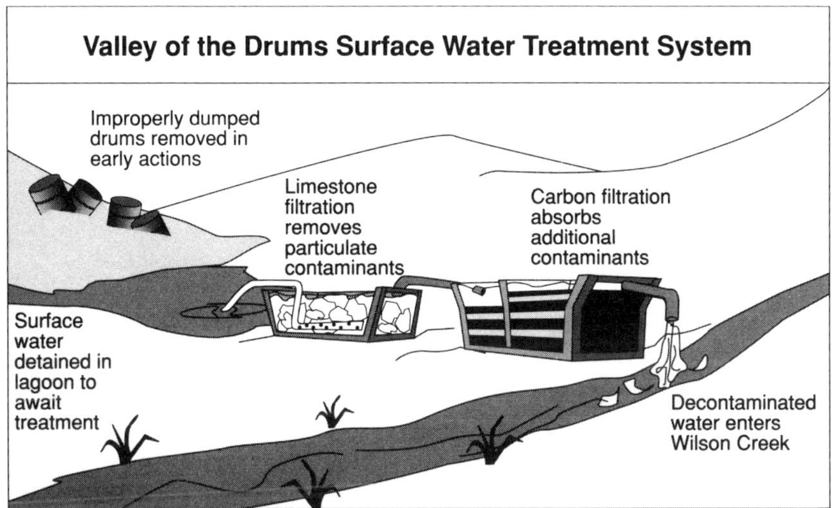

FIGURE 2.9 Vally of the Drums surface water treatment system, Bullitt County, Kentucky. [From U.S. Environmental Protection Agency (1992d).]

contaminated water from entering Wilson Creek, which flows into the Ohio River. Treatment consisted of a two-stage filter control (U.S. Environmental Protection Agency, 1992d): the first filter of crushed limestone removed particulates and the second filter used activated carbon to remove the remaining contaminants.

Another example of a Superfund site where surface water was treated is the Old Midland Products site in Yell County, Arkansas (which was described earlier in this chapter). This former wood preserving facility had lagoons containing 620,000 gallons of liquid wastes, other locations on the site containing sludges, and contaminated soils (U.S. Environmental Protection Agency, 1993e). The remediation team treated liquid wastes from the lagoons and storm water runoff from the site using an activated carbon filtration process (Fig. 2.10).

In the United States, pump and treat techniques are the most widely used for groundwater remediation. In this process, groundwater is pumped to the surface, and then contaminants are removed by a variety of methods, including air-stripping towers to remove volatile compounds, activated carbon filtration to remove toxic substances, ultraviolet or ozone treatment to oxidize organic contaminants in water, and precipitation methods to remove metals.

The Wells G&H Superfund site in Woburn, Massachusetts, used pump

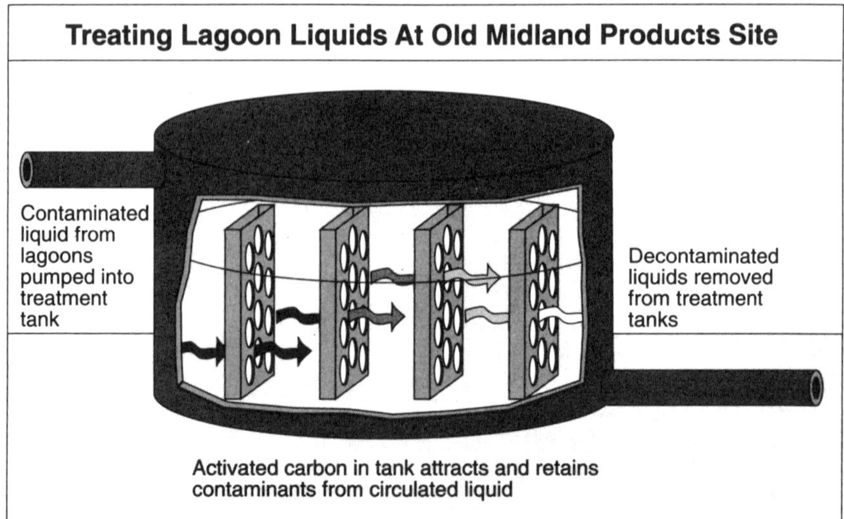

FIGURE 2.10 Treating lagoon liquids at Old Midland Products site, Yell County, Arkansas. [From U.S. Environmental Protection Agency (1993e).]

and treat techniques (see Chapter 8). A variety of VOCs contaminated this 330-acre site with two municipal drinking water wells. Investigators found about 200 damaged 55-gallon drums of industrial wastes that someone had dumped on a vacant lot near the wells (U.S. Environmental Protection Agency, 1993g). The remediation team pumped groundwater from the different wells and tested it to determine which contaminants were present. Depending on the particular mix of contaminants in the water, different treatments were used. In one innovative treatment, water was treated with ultraviolet light (UV) in combination with ozone, hydrogen peroxide, or both to oxidize the organic contaminants.

The Ciba-Geigy Superfund site in McIntosh, Alabama, is another site that is using pump and treat techniques to remediate contaminated groundwater (see Chapter 8). The 1500-acre site had produced DDT and other insecticides in the 1950s and 1960s, and these production activities had contaminated the groundwater. As part of the remediation, Ciba-Geigy installed a $3 million dollar well pumping system to capture contaminated groundwater for treatment (U.S. Environmental Protection Agency, 1992e). Ciba-Geigy treated the groundwater as well as surface runoff in a $73 million dollar wastewater treatment system that used bacteria and other biological organisms. This treatment is a form of bioremediation.

Soil vapor extraction and air-sparging techniques use pumping to remove volatile organic contaminants from the air spaces in the unsaturated zone of the soil, followed by treatment of the contaminants. Remediation at the American Thermostat Superfund site in South Cairo, New York (discussed earlier), used soil vapor extraction to remediate contaminated groundwater. By a combination of air stripping and vapor extraction, more than 10 million gallons of VOC-contaminated groundwater were treated (Fig. 2.11). Air stripping involves pumping air down a well into contaminated groundwater, whereupon VOCs that vaporize into the air bubbles are captured. For vapor extraction, the contaminated groundwater and air bubbles are then pumped out using a nearby well. Remediation at this site achieved 80 to 90% reduction in VOC concentrations in the groundwater.

Solidification and/or stabilization techniques attempt to immobilize the toxic contaminants by trapping them in a solid matrix. Vitrification stabilizes and reduces the volume of contaminated soil. It works well with inorganic compounds but not as well with organic contaminants. Long-term leaching studies and monitoring are needed before these techniques will be widely used.

However, some solidification and/or stabilization techniques are effective with certain organic chemicals. Remediation at the Pioneer Sand Company Superfund site in Pensacola, Florida, used a solidification technique to remediate a set of organic compounds known as light, nonaqueous-phase

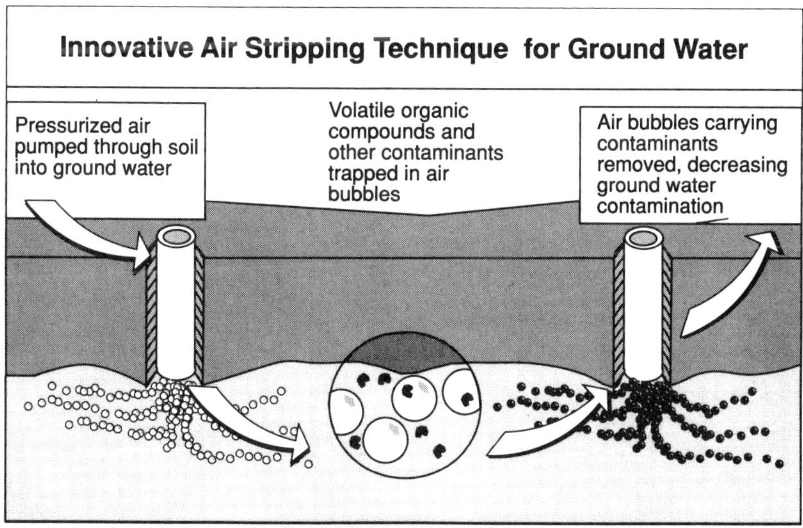

FIGURE 2.11 Innovative air-stripping technique for groundwater. [From U.S. Environmental Protection Agency (1992b).]

liquids (LNAPL). LNAPLs are undissolved chemicals, often petroleum products like gasoline and other fuels, that float on top of groundwater rather than dissolve in the groundwater. Pump and treat remediation techniques often fail to capture LNAPLs. At the Pioneer Sand Company site, pressurized air was forced down a well into the aquifer, which contained approximately 50,000 gallons of LNAPLs (U.S. Environmental Protection Agency, 1993h). The pressurized air allowed 15% of the liquid component of the LNAPLs to evaporate and caused the remaining 85% to solidify into a tarlike substance (Fig. 2.12). By solidifying the LNAPLs, this technique minimizes the possibility that the contaminants will migrate off the site or mix with the groundwater.

Remediation teams do not use chemical treatments often, but they do show promise in dechlorinating some dangerous contaminants such as PCBs.

Recycling can also be a remediation technique in special circumstances. It is possible to recycle waste when a site has a concentration of a contaminant that is sufficiently high in value to justify removing it from the site for use or sale. The Eastern Diversified Metals site in Rush Township, Pennsylvania, is an example of using recycling as a remediation technique.

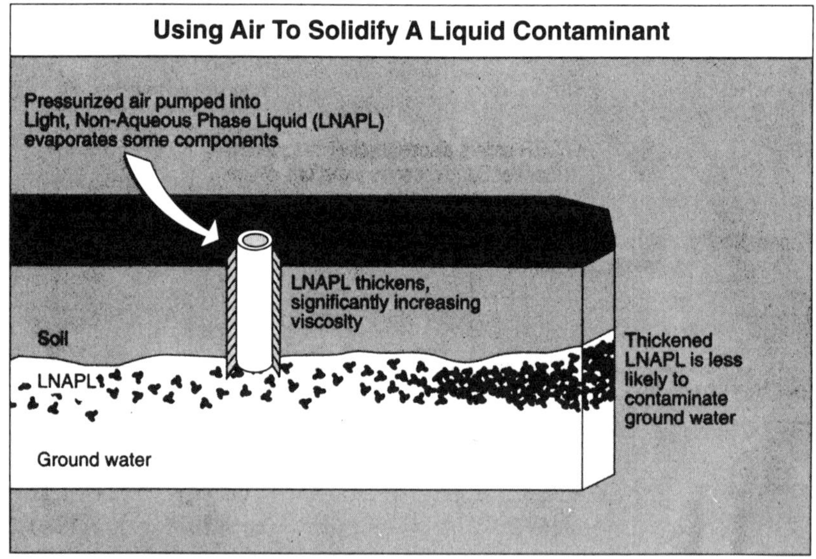

FIGURE 2.12 Using air to solidify a liquid contaminant. [From U.S. Environmental Protection Agency (1993h).]

The company operated a wire reclamation facility on a 25-acre site. From 1966 until 1977, the company stripped plastic and fiber insulation from electrical wires and cables and disposed of the insulation on a 7.5-acre portion of the site (U.S. Environmental Protection Agency, 1993i). Eventually, more than 350 million pounds of disposed insulation formed a mountainous pile of waste. Fires in the waste pile released dioxin, polychlorinated naphthalenes, PCBs, and heavy metals, including copper, lead, manganese, and zinc, into the leachate, soil, and groundwater of the site. In this case, recycling was a cost-effective cleanup technique because of the large volume of a single type of waste.

In general, two recycling methods are in use: bulk processing and sink-float processing. Bulk processing converts the waste without alteration into a solid plastic mass using heat, pressure, and some chemical additives. Companies use the resulting plastic to manufacture solid objects such as tiles, fenders, plastic lumber, and highway cones and barriers (Fig. 2.13). The second process, sink-float, adds the waste materials to a container filled with water to allow less dense components, such as plastics, to float while soil, metal, and other denser objects sink. The collected polyethylene and polyvinyl plastics become raw materials for manufacturing processes. At the Eastern Diversified Metals site, the remediation team expects that recycling will remove between 60 and 95% of the waste volume.

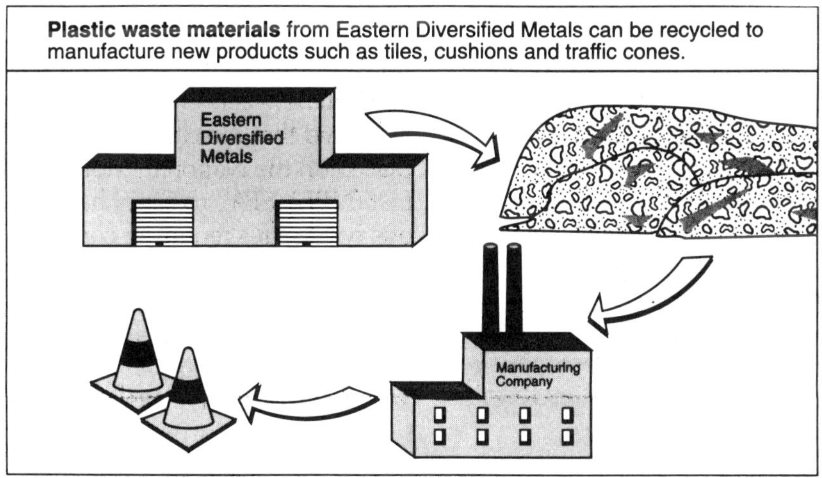

FIGURE 2.13 Recycling plastic waste materials. [From U.S. Environmental Protection Agency (1993i).]

Bioremediation Treatments

Microbial treatments to remediate toxic contamination are also known as bioremediation. Remediation teams can use these techniques in two ways: (1) after removal of contaminated soil or water or (2) without excavating or pumping the soil or water, which is called *in situ* bioremediation. They are promising methods that use natural processes, plants, and microorganisms to degrade and decrease the concentration of organic compounds on contaminated sites. Bioremediation methods attempt to maximize the rate of degradation by manipulating a number of physical, biological, and chemical parameters in the soil to either remove the chemical compound or alters its original state.

Plants may accumulate compounds in their biomass for subsequent disposal, degrade a contaminant, or allow microbes attached to the plant roots to degrade the contaminant (Bell, 1992). Microorganisms, which can be naturally occurring or introduced, degrade the contaminants to nontoxic compounds when sufficient oxygen, nitrogen, and phosphorus are present (National Research Council, 1993; Blaushild and Simon, 1993; Hinchee and Olfenbuttel, 1991).

Bioremediation techniques may be combined with other remediation technologies, either sequentially or simultaneously. Bioventing is the process of supplying oxygen to soil to stimulate biodegradation of contaminants and to promote contaminant removal by volatilization. Air sparging is the injection of pressurized air to the saturated zone, where it promotes volatilization and degradation. Passive bioremediation, also known as natural and intrinsic bioremediation, relies on unaugmented natural processes to reduce contaminant levels.

There are two discernible trends that may have a critical effect on the use of bioremediation techniques in environmental cleanup, yet it is uncertain how they will develop or if one will become dominant. One of these trends is favorable for bioremediation technologies and the other is unfavorable. How these two trends affect environmental cleanup will vary in different countries.

On the favorable side, the public views bioremediation as "natural" and thus a "green" technology, which leads to support for its expanded use for remediating toxic chemical and radionuclide contamination. In contrast, techniques that employ human intervention with chemical or physical methods are perceived as "unnatural" remediation. Bioremediation allows natural processes that rely on naturally occurring microorganisms to cleanse the earth, and so it is perceived as a human-assisted extension of the natural processes at work in the water and soil of undisturbed ecosystems. *In situ* bioremediation has the potential for remedial action without the physical disturbance that excavation and other actions require. There have been cases where the public welcomes bioremediation techniques for environmental cleanup, while being strongly opposed to incineration or chemical treatment of contaminants.

The unfavorable trend, on the other hand, is the view that bioremediation is science run amok. From this perspective, scientists are manipulating nature in potentially dangerous ways that increase the possibility of environmental disaster. The public's greatest fear is that genetically engineered microorganisms developed for bioremediation may get loose in the environment and wreak ecological havoc. An example is the possibility that a microorganism genetically altered to bioremediate petroleum contamination could get loose and destroy the planet's petroleum reserves.

Economic considerations are arguably the primary driving force in the development and use of new and innovative bioremediation techniques, for bioremediation holds the promise of significantly lower environmental cleanup costs. Yet bioremediation techniques are still under development, and remediation teams have not yet used them sufficiently to establish their costs. Initial results indicate that they hold great potential for being highly cost-effective. For organic toxic chemical contamination, it is estimated that bioremediation costs will range from one-quarter to one-half those of other remediation techniques (Snyder, 1993; De Rore, 1994).

In situ bioremediation holds the greatest potential for cost-effective environmental cleanup. Most remediation techniques require either excavating contaminated soil or pumping contaminated groundwater prior to treatment, whereas *in situ* bioremediation eliminates the substantial cost of excavating or pumping. Although the potential for *in situ* bioremediation is great, it is still unproven. Remediation teams must complete additional field trials and demonstration projects before there is sufficient experience to prove that it is cost-effective.

Even in situations where remediation teams cannot use bioremediation *in situ,* bioremediation itself may still be the most cost-effective technique. The introduction of nutrients (biostimulation) or of species of microorganisms that are known to work in concert with resident microorganisms (bioaugmentation) is often far less expensive than alternative remediation techniques. In the United States, bioremediation cleanup of soil contaminated by gasoline (refined petroleum) is becoming common. In this situation, the soil is first excavated and then nutrients are introduced to improve the speed and extent of remediation by naturally occurring microorganisms. This may be followed by the injection of microorganisms to supplement those already present.

With soil and water contaminated by gasoline, bioremediation often is more cost-effective than alternate remediation options such as incineration or air stripping. In the United States in 1993, there were 150 contaminated sites using bioremediation, most frequently for soil contamination (Table 2.1). With additional research and experience, it is likely that bioremediation techniques will prove themselves to be the most cost-effective technique available for other types of contamination, such as to destroy organic chemical solvents and to concentrate heavy metals and radionuclides.

Remediation teams are more likely to use bioremediation techniques when time is not a serious constraint. Environmental cleanups are not likely to use such techniques in the most serious cases of toxic chemical or radionuclide contamination, which almost always represent a threat to human health that requires fast action. Physical remedial actions such as excavation of contaminated soil or pumping of contaminated groundwater can more rapidly minimize the threat to public health than bioremediation techniques, which require more time to achieve cleanup. Sometimes remediation teams can follow these physical remedial actions with bioremediation, but in most instances they follow them with additional physical or chemical treatments.

Policies to promote environmental cleanup can directly affect the research and application of bioremediation techniques. For example, policy makers may craft policy that requires the use of proven technologies with good cost data. The disadvantage of such policies is that they will force countries to continue using expensive but well-known remediation techniques. Alternatively, policies can promote the use of innovative remediation techniques. It is likely that more cost-effective remediation techniques will result from policies that specifically promote innovative remediation techniques, such as bioremediation.

Policies concerning bioremediation must carefully address the issue of

TABLE 2.1

TYPE AND NUMBER OF U.S. SUPERFUND SITES USING BIOREMEDIATION IN 1993[a]

Type of media	Number of sites
Soil	111
Groundwater	58
Sediments	15
Sludge	10
Surface water	2

[a]U.S. Environmental Protection Agency (1993j).

genetic engineering. Genetic engineering techniques hold substantial potential to develop microbes that can be highly effective and inexpensive in remediating contamination. For this reason, continued research into genetic engineering for bioremediation purposes is appropriate. However, researchers must tightly control these efforts to ensure that genetically altered microbes do not escape into the environment and cause environmental damage and a potentially devastating public backlash against bioremediation.

The French Limited site in Harris County, Texas, was the first Superfund site to use bioremediation. During the 1950s and 1960s, the owners used this 22.5-acre site for sand mining, which created large pits. In 1966, the owners sold the site and the new owners used the pits for the disposal of hazardous wastes from local industries (U.S. Environmental Protection Agency, 1993k). Subsequent floods spread VOCs, phenols, heavy metals, and PCBs over the site and onto the floodplain of the nearby San Jacinto River. *In situ* bioremediation entailed pumping oxygen, nutrients, and bacteria into the contaminated soil (Fig. 2.14).

The remediation team at the Brown Wood Preserving site in Live Oak, Florida, also used bioremediation. The former wood preserving site operated from 1945 until 1978, and became a Superfund site in 1983. The company had stored the chemical sludge from wood treatment in a lagoon that periodically flooded. The sandy soil of the 55-acre site contained significant concentrations of creosote, pentachlorophenol, and polycyclic aromatic hydrocarbons (U.S. Environmental Protection Agency, 1993l). The remediation team used *in situ* bioremediation to treat the contaminated soil and other methods to remediate the 200,000 gallons of lagoon water

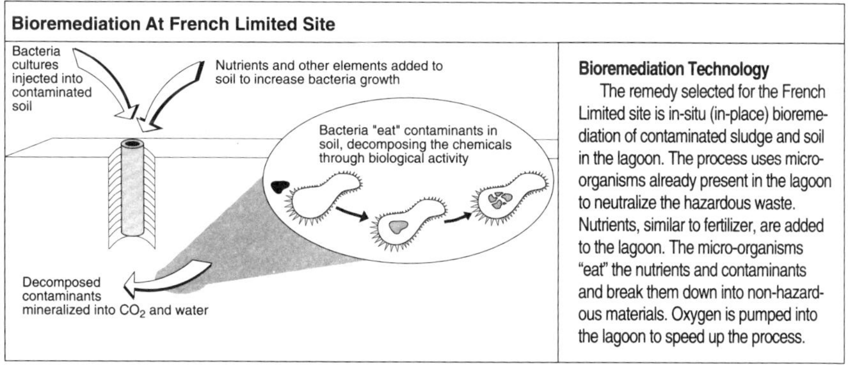

FIGURE 2.14 Bioremediation at French Limited site, Harris County, Texas. [From U.S. Environmental Protection Agency (1993k).]

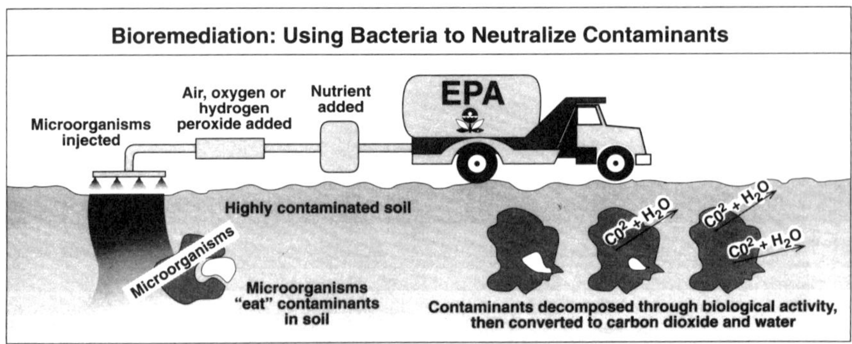

FIGURE 2.15 Bioremediation using bacteria to neutralize contaminants. [From U.S. Environmental Protection Agency (1993l).]

and 15,000 tons of creosote sediments. Bioremediation entailed injecting a mixture of nutrients, oxygen, and microorganisms into the soil (Fig. 2.15).

CONCLUSIONS

Remediation techniques are constantly improving, and policies that promote the use of innovative technologies greatly affect the development of new and more effective techniques. The U.S. Superfund program has had success in promoting innovative technologies (Table 2.2). However, it is unlikely that there is a "technological fix" to the problem of contamination and environmental cleanup. As the goals of policy aim for higher degrees of cleanup, the costs of remediation become extraordinary. In addition, some toxic contaminants have characteristics that make remediation difficult. For example, dense, nonaqueous-phase liquids (DNAPLs) migrate downward in soil and are difficult to contain or remove (Hedgcox and Stevens, 1993).

The best solution to the contamination and environmental cleanup problem is to stop using persistent toxic substances. There is considerable public support for programs calling for zero discharge, virtual elimination, or "sunsetting" of persistent toxic substances (see Chapter 4). Several countries have made progress in implementing such programs. Unfortunately, it will be a long time before we find adequate substitutes for all of these widely used substances. In the short term, the goal should be to minimize the escape of toxic substances to the natural environment and to clean up the existing contaminated sites that threaten human and ecosystem health. Countries must develop, enact, and implement effective pollution-preven-

TABLE 2.2

TYPE AND NUMBER OF U.S. SUPERFUND SITES USING INNOVATIVE TECHNOLOGIES[a]

Technology	Year of record of decision at site									
	1986	1987	1988	1989	1990	1991	1992	1993	1994	Total
Ex situ bioremediation	1	0	4	7	4	4	8	6	5	39
In situ bioremediation	0	1	2	0	3	3	4	7	5	25
Dechlorination	0	0	0	0	1	2	0	0	0	3
In situ flushing	1	0	1	3	1	3	4	2	3	18
In situ vitrification	0	0	0	0	0	1	0	0	0	1
Soil washing	0	0	2	2	6	1	1	0	0	12
Solvent extraction	0	0	0	3	0	1	0	1	1	6
Thermal desorption	1	3	4	2	8	10	5	10	7	50
Soil vapor extraction	2	1	7	22	17	33	18	22	11	133
Other technologies	0	0	0	0	0	1	2	0	1	4
Total	5	5	20	39	40	59	42	48	33	291

[a]Source: U.S. General Accounting Office (1996b).

tion and remediation policies if they are to solve the problem of toxic contamination. Remediation techniques are available, but most countries fail to remediate many of their contaminated sites. Some countries have systems to identify the most seriously contaminated sites and to remediate these properties first. The following chapters present different aspects of the problem and policy options to contend with the contamination and environmental cleanup problem.

References

Bell, R. M. 1992. *Higher Plant Accumulation of Organic Pollutants from Soils* (PB92-209 378/AS). Springfield, VA: National Technical Information Service.

Blaushild, D. I., and J. A. Simon. 1993. Recent developments in cleanup technologies. *Remediation* **3**(4), 495–501.

Clark, R. M. 1987. Technological approaches to removing toxic contaminants, pp. 89–124. In G. W. Page (ed.), *Planning for Groundwater Protection.* Orlando, FL: Academic Press.

Cohen, R. M., R. R. Rabold, C. R. Fraust, J. O. Rumbaugh, III, and J. R. Bridge. 1987. Investigation and hydraulic containment of chemical migration: Four landfills in Niagara Falls. *Civ. Eng. Pract., J. Boston Soc. Civ. Eng., Sect./Am. Soc. Civ. Eng.* **2**(1), 33–58.

De Rore, H. 1994. Biotechnological processes for cleaning soils and sediments polluted with organics. Presentation at the NATO Advanced Research Workshop, Biotechnologies for Radioactive and Toxic Wastes Management and Site Restoration: Scientific, Educational, Business Aspects, Mol, Belgium, November 28–December 2.

Hedgcox, H. R., and W. S. Stevens, 1993. An assessment of hydraulic containment to control vertical DNAPL migration. *Remediation* **3**(4), 443–458.

Hinchee, R. E., and R. F. Olfenbuttel (eds.). 1991. *In Situ Bioreclamation.* Boston: Butterworth-Heinemann.

Kinner, L. L., A. McGowin, S. E. Manahan, and D. W. Larsen, 1993. Reverse-burn gasification for treatment of hazardous wastes: Contaminated soil, mixed wastes, and spent activated carbon generation. *Environ. Sci. Technol.* **27**(3), 482–488.

McKay, D. M., and J. A. Cherry. 1989. Groundwater contamination: Pump-and treat remediation. *Environ. Sci. Technol.* **23**(6), 620–636.

National Research Council. 1993. *In Situ Bioremediation: When Does It Work?* Washington, DC: National Academy Press.

Page, G. W. 1987a. Wausau, Wisconsin, case study, pp. 241–260. In G. W. Page (ed.), *Planning for Groundwater Protection.* Orlando, FL: Academic Press.

Page, G. W. 1987b. South Brunswick, New Jersey, case study, pp. 325–340. In G. W. Page (ed.), *Planning for Groundwater Protection.* Orlando, FL: Academic Press.

Rose, D. 1994. DOD, DOE clean-up efforts promising. *Environ. Prot.* **5**(1), 58–64.

Snyder, J. D. 1993. Off-the-shelf bugs hungrily gobble our nastiest pollutants. *Smithsonian* **24**(1), 66–76.

U.S. Environmental Protection Agency. 1990. *Summary of Treatment Technology Effectiveness for Contaminated Soil,* (PB92-963 351). Emergency and Remedial Response. Washington, DC: U.S. Government Printing Office.

U.S. Environmental Protection Agency. 1992a. *Superfund at Work. EPA and Polluter Work Together to Clean Superfund Site in North Carolina* (EPA 520/F-92-005). Washington, DC.

U.S. Environmental Protection Agency. 1992b. *Superfund at Work. EPA Mobilizes to Safeguard Community and Eliminate Chemical Waste* (EPA 520/ F-92-009). Washington, DC.

U.S. Environmental Protection Agency, 1992c. *Superfund at Work. EPA Uses Innovative Technology to Eliminate Wide Beach PCB Threat* (EPA 520/ F-92-010). Solid Waste and Emergency Response. Washington, DC.

U.S. Environmental Protection Agency. 1992d. *Superfund at Work. Valley of the Drums Cleanup: A Superfund Benchmark* (EPA 520/F-92-006). Solid Waste and Emergency Response. Washington, DC.

U.S. Environmental Protection Agency. 1992e. *Superfund at Work. EPA Wins Ciba-Geigy's Full Cooperation to Clean Up Alabama Site* (EPA 520/ F-92-017). Solid Waste and Emergency Response, Washington, DC.

U.S. Environmental Protection Agency. 1993a. *Superfund at Work. Pesticide Contamination Addressed at Powersville Dump Site* (EPA 520/F-93-005). Washington, DC.

U.S. Environmental Protection Agency. 1993b. *Superfund at Work. EPA Actions Abate Community Exposure to Radium Facility* (EPA 520/F-93-006). Washington, DC.

U.S. Environmental Protection Agency. 1993c. *Superfund at Work. Cleanup Nearly Complete at Harvey and Knott Site* (EPA 520/F-93-007). Washington, DC.

U.S. Environmental Protection Agency. 1993d. *Superfund at Work. EPA Orders Incineration of Hazardous Chemicals* (EPA 520/F-94-003). Washington, DC.

U.S. Environmental Protection Agency. 1993e. *Superfund at Work. Superfund Site Clean Enough for Unrestricted Use?* (EPA 520/F-93-003). Washington, DC.

U.S. Environmental Protection Agency. 1993f. *Superfund at Work. Hazardous Waste Contributors Pay for Fulton Terminals Site Cleanup* (EPA 520/ F-93-008). Washington, DC.

U.S. Environmental Protection Agency. 1993g. *Superfund at Work. Cleanup Begins at Wells G&H, One Year after Landmark New England Settlement* (EPA 520/F-93-015). Washington, DC.

U.S. Environmental Protection Agency. 1993h. *Superfund at Work. EPA Oversees Cleanup of Pioneer Sand Company Site* (EPA 520/F-93-012). Washington, DC.

U.S. Environmental Protection Agency. 1993i. *Superfund at Work. EPA Pioneers Cleanup Strategy: Recycling Hazardous Waste* (EPA 520/F-93-0018). Washington, DC.

U.S. Environmental Protection Agency. 1993j. *Bioremediation: Innovative Pollution Treatment Technology* (EPA 640/K-93-002). Office of Research and Development. Washington, DC.

U.S. Environmental Protection Agency. 1993k. *Superfund at Work. Innovative Technology Used to Clean Up French Limited* (EPA 520/F-93-004). Washington, DC.

U.S. Environmental Protection Agency. 1993l. *Superfund at Work. Innovative Technology Used to Restore Environment* (EPA 520/F-94-001). Washington, DC.

U.S. Environmental Protection Agency. 1994a. *Superfund at Work. EPA Orders Incineration of Hazardous Chemicals* (EPA 520/F-94-003). Washington, DC.

U.S. Environmental Protection Agency. 1994b. *Superfund at Work. Superfund Tackles Operation that Spawned Four Waste Sites* (EPA 520/F-94-011). Washington, DC.

U.S. Environmental Protection Agency. 1994c. *Superfund at Work. Enforcement Tools Allocate Liability, Speed Cleanup* (EPA 520/F-94-012). Washington, DC.

U.S. Environmental Protection Agency. 1995. *Superfund at Work. Former Metals Reclamation Facility to be "Delisted"* (EPA 520/F-95-006). Washington, DC.

U.S. General Accounting Office. 1991. *Hazardous Waste: Limited Progress in Closing and Cleaning Up Contaminated Facilities* (GAO/RCED-91-79). Washington, DC: U.S. Government Printing Office.

U.S. General Accounting Office. 1995. *Hazardous Waste: Compliance with Groundwater Monitoring Requirements at Land Disposal Facilities* (GAO/ RCED-95-75BR). Washington, DC: U.S. Government Printing Office.

U.S. General Accounting Office. 1996a. *Superfund: EPA Has Identified Limited Alternatives to Incineration for Cleaning Up PCB and Dioxin Contamination* (GAO/RCED-96-13). Washington, DC: U.S. Government Printing Office.

U.S. General Accounting Office: 1996b. *Superfund: Use of Innovative Technologies for Site Cleanup* (GAO/T-RCED-96-45). Washington, DC: U.S. Government Printing Office.

CHAPTER 3

PUBLIC HEALTH IMPLICATIONS OF CONTAMINATED SITES AND ENVIRONMENTAL CLEANUP

INTRODUCTION

The primary motivation for remediating contaminated sites is to protect human health. Toxic contaminants also threaten the health of exposed plants, animals, and ecosystems and negatively affect the economic and social circumstances of individuals, firms, and communities. However, it is the threat to human health that mobilizes political forces that seek a solution to environmental contamination.

Contaminated sites are a human health threat because many of the contaminants present in the soil or groundwater are capable of producing human morbidity (illness) and mortality (death) after exposure to even minute concentrations. Furthermore, many toxic substances are highly persistent in the environment and therefore contaminated sites may be a public health threat far into the future unless remediation takes place. And because contaminated sites often are not known and publicly identified, people may live, work, or play on them without knowledge of the potential health risk.

Exceptionally low concentrations of toxic substances can threaten health. The human senses of smell, sight, and taste usually are not able to sense the exposure to toxins at such concentrations, and evidence of a health threat from contaminated land may not be evident for long periods after exposure has begun. Humans who are exposed to toxic substances may not manifest observable morbidity, mortality, or genetic damage until decades after the exposure.

Scientists have well-established evidence of the toxicity of many contaminants. Yet, even though there is indirect evidence that contaminated sites may be a significant health threat, there is little direct evidence. In many cases, the pathways by which toxins from contaminated sites come into contact with humans are known. Changes in long-term health trends suggest that toxic chemical compounds may be influencing human health, especially via cancers. The psychological factors associated with involuntary exposure to contaminants can produce health risks in addition to those caused by actual exposure. Despite methodological shortcomings, there have been studies that suggest a relationship between contamination and adverse health. Taken together, these factors have raised concerns in the public health community and have produced fear, political mobilization, and direct action among the general public.

In several countries there is widespread public fear about the health threat caused by toxic contaminants. The media have often publicized incidents of problems associated with contaminated sites, using terms such as "ticking toxic time bomb" to describe them. There is some evidence that the public perception of the health risk caused by contaminated sites far exceeds the actual health risk.

EXPOSURE

Contaminated sites have toxic substances in surface water, the soil, or groundwater that are not securely confined. The fact that toxic contaminants are free to follow environmental pathways suggests that contaminated sites are contributing to the quantity of toxic substances loose in the environment. Human exposure to these substances may cause negative health effects.

The exposure pathways are a critical link in the contamination and environmental cleanup problem. Exposure is difficult to predict because environmental pathways that may allow the movement of toxic contaminants are site specific and laborious to identify and measure (U.S. Environmental Protection Agency, 1977).

Exposure Pathways

There are several ways that people may get exposed to toxic substances found on contaminated sites. Direct poisoning is possible when people have access to the contaminated land, in which case exposure may occur by ingestion, inhalation, or dermal (through the skin) absorption from contact with contaminated soil or water. Derelict land in urban areas is often contaminated, usually by former uses of the site or illegal "midnight dumping" of toxic wastes (Page, 1987a). Open spaces in densely populated urban areas often attract neighborhood children, the homeless, people walking dogs, and others who may be exposed to toxins on the site. Private firms that still own contaminated sites usually fence the property to keep people out and thereby minimize their liability for any adverse health effects to people exposed on the site. Derelict contaminated sites often do not have fences or the fences have gaps. There have been cases where the local government knows of a contaminated derelict site but will not fence the property (see Chapter 5). In the United States, municipalities may refuse to erect fences to avoid active management of the property, because such action could make them liable under CERCLA legislation (Page and Rabinowitz, 1993).

Airborne toxic substances easily come into contact with people. Fires or explosions are the most likely sources of toxic air pollutants and can result in dangerous levels of toxins in the atmosphere. Though most governments make open burning of hazardous wastes illegal, fires and explosions do occur. They are especially prevalent when there are structures on the contaminated site and if the site is no longer in productive use. Gases may migrate through the soil and collect in basements, pipes, or other structures, where exposure or explosion may take place. Volatile toxic contaminants may escape to the atmosphere and cause exposure, although rapid diffusion in the atmosphere may limit high levels of exposure at large distances from the source.

Water on the surface of a contaminated site or flowing off the property, such as ponds, lakes, wetlands, and streams, may be a source of exposure (Wallach and Shabtai, 1993). Sometimes these water bodies extend beyond the boundaries of the contaminated property and people may be unaware of the upstream sources of the water. Snowmelt or precipitation events may cause water to flow off the contaminated site as surface runoff. In most cases, trace concentrations of contaminants would be a threat only if consumed. Higher concentrations of toxic substances in water could be a health threat from contact.

Exposure to contaminated groundwater holds the greatest potential risk of large-scale human exposure. Toxic contaminants released to the soil often percolate into groundwater and, once there, plumes of contaminants may migrate in directions and at speeds that are difficult to predict. Contaminants flowing in groundwater often remain discrete plumes with high concentrations of contaminants (M. P. Anderson, 1987; Office of Technology Assessment, 1984; Page, 1981). Dangerous exposure can result when a plume of contaminated groundwater intersects a drinking water well. A contaminated public water supply well can expose large numbers of people.

Contaminated sites may also cause indirect poisoning via the food chain. Eating plants or animals that contain toxic substances from a contaminated site can expose people to health risks. Many contaminants are fat soluble and bioaccumulate in the food chain. The food chain is the most likely exposure route for the least affluent segments of some societies, who may retain strong ties to a rural past or are recent immigrants for whom hunting, fishing, and collecting wild edibles remain part of their lifestyle. These segments of society have low incomes that make these non-cash-economy sources of food attractive. In addition, they may be unaware of the potential routes of exposure or of the health threats caused by exposure to toxic contaminants in the environment and especially via the food chain.

Exposure Evidence

Scientists have evidence that groups of people have come in contact with toxic substances from contaminated sites. The instances of high levels of exposure are small in number but serious. The evidence of chronic low-level exposure is also clear, but the human health implications are controversial. The exposure pathways may be the same in both cases, but the concentration of contaminant to which people are exposed can differ by many orders of magnitude.

High levels of exposure to toxic substances from contaminated sites most often occur directly on the contaminated property. Some hazardous waste disposal sites were little more than open dump sites, and prior to policies that prevented toxic contamination, chemical analyses of ponds or streams often revealed high concentrations of toxic substances. On many sites, investigators found combinations of full, leaking, and empty storage drums of liquid wastes. Chemical analysis showed them to hold toxic substances; often these drums had labels correctly stating their toxic contents. Leaking transmission pipelines and storage tanks often leaked large volumes of liquid toxic substances that contaminated soil and groundwater sources of drinking water. In the economically advanced countries, standards for pipes, tanks, and monitoring systems have dramatically reduced, but not eliminated, these sources of gross contamination. Case studies of contaminated sites causing high levels of human exposure to toxic substances reveal that serious health risks can occur in locations that are difficult to predict (Dienemann *et al.*, 1991; Page, 1987b).

There are many studies of exposure and health risk from Superfund sites in the United States. In a study of 50 Superfund sites in the Great Lakes region, researchers judged 68% of the sites to be of indeterminate/potential health risk, 12% as no health risk, and 20% as having a known health risk associated with exposure at the site (Myers *et al.*, 1995).

Even though exposures to high levels of toxic substances are clearly dangerous, the public health threat of chronic exposure to low levels of exposure is controversial. Technical advances in analytic chemistry, especially instrumentation, starting in the 1970s have enabled scientists to determine that people are exposed to incredibly small concentrations of toxic substances. Chemical analyses reveal that virtually all humans have measurable concentrations of toxins, such as polychlorinated biphenyls, chlorinated pesticides, and heavy metals, in their bodies. One U.S. study found trace concentrations of more than 200 industrial chemicals in the population sampled (Federal Emergency Management Agency, 1990, p. 1–2).

Our ability to measure exposure exceeds the ability of environmental health scientists to interpret the significance of these low exposures. We do not know the human health effects of exposure at concentrations in the low parts per billion or parts per trillion range (Page, 1987c; Greenberg and Page, 1981). This is known as the problem of the vanishing zero. The toxic substances that were once reported as nondetectable or zero are now reported to be present at microcontaminant concentrations. Scientists, public officials, and the public must decide what level of risk from exposure is acceptable.

TOXICITY AND HUMAN HEALTH EFFECTS

The word "toxic" means poisonous. How toxic are environmental contaminants and what are the potential human health effects of exposure to them? It is essential that we understand the limits of our toxicological knowledge of contaminants, especially at microcontaminant concentrations. The toxicity at contaminated sites is critical to several questions: How contaminated must the site be to require remediation, and what levels of contamination may safely remain on the land after remediation is completed?

It is surprising how little we know about the toxicity of most of the chemical compounds found on contaminated sites and about their effects on human health. The chemical revolution of the past several decades has created millions of synthetic organic compounds that had previously never existed. Scientists have not tested most of these new compounds to determine their toxicity, although they have identified and tested the toxicity of the most common toxic substances that contaminate land. However, we have a poor understanding of the synergistic effects of exposure to two or more toxins and of multiple chemical sensitivity.

We know quite a lot about the "acute" toxicity of many contaminants. Acute toxicity refers to toxic effects that are intense and immediate and closely correspond to what we normally think of as instances of poisoning. Acute toxic effects occur only after relatively high levels of exposure. Our knowledge of acute effects is derived from occupational exposure rather than exposure in the outdoor environment. Studies of occupational exposure to many toxic substances report potentially serious health effects, but can identify only a small number of risks because of limitations with the methods. Limited numbers of studies exist because few workers develop obvious symptoms and exposed workers may change employment or die before a health effect is apparent. There are only rough estimates of how much exposure occurred, because most occupational studies are retrospective, meaning they are conducted long after the exposure took place. Furthermore, it is difficult to control for genetic predisposition or life-style factors.

Epidemiological studies are the primary research technique that scientists use to determine if toxins from contaminated sites are causing significant increases in human disease. Most studies of occupational health are epidemiological studies. Several types of such studies are used to assess the risk of exposure to toxins in the environment (Ricci and Molton, 1985): descriptive epidemiological studies that search for patterns in human health data that may suggest factors that warrant further investigation, and analytic epidemiological studies that test hypotheses using controlled experiments.

Epidemiological studies investigating the health effects of contaminated sites have shown mixed results for several reasons. It is difficult to prove that past exposure caused the observed effects, and these effects have many possible causes in a mobile population. Most of the studies suggesting a link between contaminated sites and adverse human health are inferential and had flaws in their methods. The methodological difficulties, long time span, and high cost of a well-controlled analytic epidemiological study of contaminated sites and cancers have thus limited the results available. Although there are valid reasons for concern, evidence that contaminated sites cause adverse human health effects is not clear (U.S. National Research Council, 1991; R. F. Anderson, 1987; Pigen, 1984). Nonetheless, there is wide agreement that substantial risks can result from exposure to contaminated sites (U.S. Congress, 1989, p. 22).

Most of the toxins from contaminated sites expose people at low levels, usually in the microcontaminant range, and the toxic effects of such exposure are not clear. Many diseases, with the notable exception of cancers, have proven thresholds of a determined dose below which there is zero probability of producing an adverse health effect (U.S. National Academy of Sciences, 1977). Some scientists argue that some exposure to toxins from contaminated sites is so infinitesimal that they have no effect. The majority belief, however, is that any exposure to a carcinogen entails some risk. In the United States, government policy has dictated that policies build in a conservative margin of safety to protect people from chronic exposure to carcinogens.

Scientists estimate the cancer risk of exposure to low concentrations of environmental contaminants based on a combination of human health data and animal studies. The human health data are primarily from occupational health and other epidemiological studies. Animal studies, on the other hand, divide the animals into cases and controls, and then expose the cases to increasing doses of the chemical compound being studied. When a predetermined percentage of the animals die, the scientists autopsy the animals and estimate what levels of exposure produce diseases.

There is criticism of using animal studies to determine toxicity, for some compounds produce different effects in different species. Although some species respond differently from humans, all known carcinogens in humans, with the possible exception of arsenic and benzene, also act as carcinogens in some animal species (U.S. National Academy of Sciences, 1977). The large doses of the chemical compound used in animal studies (maximum tolerated dose), the use of solvents, and use of animals genetically bred to be susceptible to tumors are also controversial. Scientists must design animal studies to detect both toxic effects of exposure over a lifetime and the level of exposure capable of producing cancers in the most susceptible segment of the population. Conducting studies with a high number of test animals is difficult. When a chemical compound causes cancer only in the most susceptible individuals, studies can be exorbitantly expensive. For expediency and cost reasons, scientists resort to animal test trials of a small number of selected animals exposed to higher concentrations of the chemical compound (U.S. National Research Council, 1993). Despite these steps, animal testing for a compound can easily cost one million U.S. dollars.

Short-term toxicity tests are available that use bacteria and other lower organisms to test for mutagenic properties. When bacteria have mutagenic responses to exposure to a compound, it indicates that the compound may have mutagenic, teratogenic, or carcinogenic properties in humans. The Ames test is the most famous of these tests, which are fast and inexpensive compared to animal studies.

The most difficult part of determining the toxicity of environmental contaminants is estimating the lifetime human health effects of exposure to low concentrations of toxins. As noted earlier, this is the most common type of exposure from contaminated sites. Scientists make these risk estimates by using a variety of dose–response functions, which are mathematical extrapolations based on available data (Ricci and Molton, 1985; U.S. National Academy of Sciences, 1980a,b). We rely upon this information to decide when land is so contaminated that remediation is required and how clean the land must be after remediation.

There is much that we still do not know about toxicity and human health effects, however, in the past two decades we have learned a great deal. Many toxic substances can be both acutely and chronically toxic; they can have neurological, behavioral, or immunological effects; and they can be teratogenic, mutagenic, or carcinogenic. Some toxic substances are neurotoxins that attack nerve cells and produce neurological and other behavioral disorders, and others produce immunological effects, meaning they impede the body's ability to develop an immune response when exposed to some disease-causing agent. Teratogenic effects produce malformations in the fetus, mutagenic effects induce genetic changes that may produce health problems in future generations, and carcinogenic effects result in cancer.

Numerous reports suggest that toxins from contaminated sites are causing adverse human health effects, but most suffer from methodological difficulties. There is clear evidence that fish and birds exposed to toxins from contaminated sites develop disease, tumors, and deformities, yet studies of humans living near contaminated sites are less conclusive. Often, the population exposed is usually too small to conduct a statistically valid study, and the lag period between exposure and onset of disease is considerable, especially for the cancers that are often the primary concern.

Human health effects data is often suspect because of the organization collecting it or the methods they use. People living near contaminated sites often conduct a survey of neighbors to find any evidence of an elevated rate of disease. They suspect that the toxins from the nearby contaminated site cause elevated rates for certain diseases. Such findings of local disease are called "disease clusters." These studies usually suggest that the toxins on the contaminated site caused the cluster, though there are serious difficulties in scientifically investigating a disease cluster and its apparent environmental cause. Few follow-up studies of disease clusters have proved that toxins from contaminated sites are the primary cause (Wartenberg and Greenberg, 1992).

RISK PERCEPTION AND PSYCHOLOGICAL RISK

The general public perceives the presence or potential presence of contaminated sites as a risk. The media provide extensive coverage of instances of contaminated sites because the public is intensely interested in this problem. Public perceptions and fears of health risk are often greater than are warranted by the empirical evidence of risk.

Exposure to toxins from contaminated sites is an involuntary exposure to a health risk, and such exposures cause greater stress and fear than voluntary exposures of equal or even greater risk to health (Slovic *et al.*, 1979). Life entails taking some risks, and most people risk their health on a daily basis when they walk across a street, drive an automobile, or consume foods high in fats. Large numbers of people smoke tobacco despite well-publicized reports claiming that 80–90% of chronic lung diseases are caused by smoking. Likewise, many people drive a motor vehicle soon after consuming alcohol despite evidence that 50% of accidents involve alcohol use. Yet these individuals are likely to become psychologically stressed, frightened, and angry if they are involuntarily forced into exposure to a health risk, even if that risk is small in comparison to risks to which they voluntarily expose themselves. There is evidence that psychological stress from involuntary exposure by itself can be a serious health risk (Calkins, 1987). Involuntary technological risks, such as those associated with contaminated sites, are more stressful than natural hazards such as floods, earthquakes, or hurricanes. People accept natural hazards as uncontrollable, whereas people are frustrated when forced to accept technological hazards that are considered evidence of human failure (Baum *et al.*, 1983).

Involuntary technological risks, such as living near a contaminated site, can produce emotions strong enough to induce people to engage in direct action. Greenberg and Anderson (1984) summarized incidents in which a heterogeneous cross section of American society committed acts of violence, sabotage, and general civil disobedience related to new or existing hazardous waste disposal sites. Fears and concerns over the Love Canal site in the United States led to citizens kidnapping an EPA official to dramatize the community's concerns (Gibbs, 1982).

Such fear and stress make living or working near contaminated sites potent emotional and political issues. They are responsible for the NIMBY effect ("not in my backyard"), which makes the siting of new waste disposal and other activities that pose perceived environmental risks exceedingly difficult (Williard and Swenson, 1984). The resulting heightened environmental awareness can have beneficial effects, such as strong support for environmental cleanup policies. However, fear sometimes prevents the use of on-site remediation techniques, especially incineration, that would reduce the health threat of toxins on the contaminated site. Scientists have devoted considerable attention to potential techniques to overcome fear and psychological stresses concerning potential risks of toxic exposure, yet these efforts have produced few solutions (Hirshorn, 1984; Elliot, 1984).

U.S. legislation requires the Centers for Disease Control to maintain a registry of persons exposed to toxic substances near some remediation sites and to conduct psychological impact investigations, among other tests (R. F. Anderson, 1987).

The public is often alarmed and upset that their fears and health concerns do not receive sufficient attention by public officials. In the United States, citizens produce more than 1000 disease cluster reports each year, although the government discourages these citizen reports (Greenberg and Wartenberg, 1992). Responding to citizen reports of disease clusters is a difficult issue because local health departments rarely have the resources or the expertise to respond adequately. Nearly half the states have never conducted a field investigation requiring data collection of a disease cluster, and the overall field investigation rate is only 1–3% of the reports received (Wartenberg and Greenberg, 1992). Lack of official action in response to disease cluster reports often increases fear and anger among citizens living near contaminated land.

Public risk perception is a major ingredient in the contamination and environmental cleanup problem. The fear of contaminated sites is strong and the fear in itself may sometimes become a public health risk. An aroused public can be helpful in galvanizing support for efforts to contend with contaminated sites, but an aroused public can also hinder efforts to investigate and ameliorate such sites. The perception of health risk caused by contamination and environmental cleanup can reach hysterical levels that significantly interfere with decision making and remediation efforts for specific contaminated properties. Increased public education about the health risks of contaminated sites and greater opportunities for public participation in decisions about environmental cleanup may help bring public fear in line with public risk.

RISK ASSESSMENT

Risk assessment is the process of identifying risks, understanding risks, and estimating their relative magnitudes. It is a crude and unusually biased technique that is more art than science, and there is often little agreement among the risk assessment community about the methods used or the validity of the results. Nonetheless, risk assessment is a valuable analytic technique for systematically evaluating different environmental and public health risks and their relative importance and for advising policy makers how to allocate limited resources to produce the greatest risk reduction.

The methods used to measure or estimate potential exposure to toxins from contaminated sites, called "risk assessment" techniques, usually do not produce definitive results. The risk assessment of exposure to toxins from contaminated sites typically yields imprecise estimates of exposure and risk that can vary widely depending on who performs the risk assessment. Often these risk assessments vary by a factor of 100 or more (U.S. Congress, 1989, p. 22). In risk assessments of Superfund sites in the United States, it is common to assess the hypothetical exposure that may occur at a site as a form of worst-case scenario. Some hypothetical assessments have concluded that the contamination was so severe that 70% of the exposed population would die of cancer (U.S. Congress, 1989, p. 22).

Risk assessment is a mechanism for assembling and synthesizing relevant information on risks for a particular site. The Superfund program requires a risk assessment for each contaminated site that EPA evaluates for inclusion in the program (U.S. General Accounting Office, 1991). The risk assessment calculates the risks that sites pose to human health if left untreated and it helps determine the extent of remediation needed to avoid adverse health effects. It combines two components to estimate the likelihood that individuals at or near a contaminated site could develop cancer or other health problems: the exposure assessment and the toxicity assessment. The exposure assessment estimates the quantity of contaminants likely to reach people at or near contaminated sites, and the toxicity assessment estimates the contaminant intake level that could harm human health. The risk of cancer is usually the most significant risk when conducting risk assessments of contaminated sites. Total cancer risk is the sum of the reasonable maximum exposure for each contaminant across each possible pathway exposure. The procedures and the outcomes of both exposure and toxicity assessments for contaminated sites are often controversial.

The exposure assessment is controversial because risk assessors often lack solid field data and are forced to make assumptions about potential exposure pathways. Detailed surveys of people living near contaminated sites for site-specific exposure data are usually limited because of their difficulty, time constraints, and expense. It is not easy to identify all of the contaminants in the body tissues of people potentially affected by a contaminated site and then to determine the source of those contaminants. Surveys can discover how people currently use a contaminated site, areas nearby, how much water people drink daily, and other life-style factors. However, it is usually impossible to discover definitive answers to important issues. Some of the most difficult questions are how long people will live near the site and how land on the site and nearby will be used in the future.

Assumptions about the future use of contaminated sites and nearby land are critical steps in estimating future exposures to contaminants and therefore to determining a strategy for environmental cleanup. Countries need more stringent cleanup for sites on which people will reside in the future than for sites that people will use for business activities. Residential use of land implies a greater number of exposure pathways, higher levels of water consumption, and a longer period of exposure. Some countries (e.g., the United Kingdom, see Chapter 11) explicitly factor in future intended land use when making decisions about environmental cleanup of contaminated land. Other countries want all land cleaned up to the highest standards (e.g., The Netherlands, see Chapter 10). The U.S. Superfund program generally assumes that residential use of contaminated sites is possible in the future, unless there is substantial contrary evidence.

Analysts performing risk assessment of contaminated sites often must use standardized exposure assumptions to estimate how contaminants from a site might reach people. These assumptions typically include information on the length of time people are likely to live in their present residence and how much water they consume in a day. In the United States, the Superfund program specifies that analysts must use the following conservative assumptions: that people consume two liters of water daily, which would be a potential exposure pathway, and that people live in their homes for 30 years. Critics argue that these assumptions are too conservative. Surveys report that only 10% of Americans drink more than two liters of water per day at home and that only 7% live in their homes longer than 30 years (Guerrero, 1995). Note that while critics argue this approach is too conservative, as many as one in ten people near a site may receive greater exposure than is estimated.

Governments often make conservative assumptions to protect the public. The U.S. EPA uses these assumptions because it claims a mandate to protect children and susceptible people from the risks of contaminated sites, not just the average American. However, critics claim that the cumulative effect of a series of conservative assumptions in the risk assessment process produces extreme results. For example, the Superfund program requires a risk assessment for each suspected carcinogen present at the site, and these risks are then summed to produce the cumulative risk assessment. The sum of these conservative individual risks (only a 5% chance that the actual risk is lower than the reported risk) creates an even more conservative total risk for the Superfund site. If dozens of carcinogens are present at a site, the cumulative impact can be significant.

Some analysts propose that risk assessment should have a more central role in the Superfund program. At present, Superfund sometimes uses risk assessment to help establish targets for how clean the site must be after cleanup. EPA uses risk assessment for sites where "applicable or relevant and appropriate requirements" (ARARs) do not apply (see Chapter 8), although often in conjunction with ARARs at sites where ARARs do apply. However, there is no specific guidance for using site-specific risk assessment procedures as a tool to establish cleanup levels or to integrate the use of risk assessment with ARARs. Some analysts propose that EPA develop standard risk assessment procedures and exposure scenarios that include explicit consideration of a site's prospective use (Clean Sites, Inc., 1990). They advocate that approach to provide consistency in cases where standards do not exist.

The U.S. government intends that the Superfund program should clean up only the most serious cases of contaminated sites. For example, in Broward County, Florida, in 1994, there were 979 identified contaminated sites, yet only six (0.006%) were on the National Priority List of the Superfund program (Florida Department of Natural Resources Protection, 1994).

Risk assessments suggest that many Superfund sites probably present a relatively small environmental and public health threat. They are modest even when compared to other environmental problems such as radon, lead, and asbestos contamination that also require remedial actions. Experts rated risks of Superfund and Resource Conservation and Recovery Act (RCRA) sites as moderate to low, whereas they rated the risks of indoor radon, other indoor air pollution, ozone depletion, and global warming as high (U.S. Environmental Protection Agency, 1987). Superfund risks rated eighth out of 29 environmental risks that the analysts evaluated in terms of cancer risk, relatively low in noncancer health risks, medium in ecological risks, and medium in welfare effects (U.S. Environmental Protection Agency, 1987).

The actual human health risks that the aggregate contaminated sites represent are still unknown. A committee of the U.S. National Research Council reviewed the literature and could not confirm or refute either the public's opinion that contaminated sites represent a great health risk or the risk assessment view that they represent a small risk (U.S. National Research Council, 1991). Insufficient information about the connection between exposure and effect precludes clear assessment of the actual health risk of contaminated sites.

Despite considerable scientific uncertainty, the risk assessment process does estimate the human health risks of individual contaminated sites. An assessment of the magnitude of risk at each site is critical in deciding whether to remediate the contamination and how much remediation is necessary. Countries have used two approaches concerning risk: (1) risk must be reduced to a level suitable for any future use of the site (for examples of this multifunctionality approach see Chapters 8 and 10) and (2) remediation must reduce risk to a level suitable for the specific future use proposed for the site (see Chapter 11). Both approaches must estimate risks and make a decision about the level of risk that will exist after the environmental cleanup. The Superfund program uses a standard of one in a million incremental deaths from cancer as the starting point in decisions about the level of risk acceptable after environmental cleanup. All Superfund decisions fall within the range of 10^{-4} to 10^{-6} incremental lifetime cancer risk. Risk analysts combine cancer risks with noncancer health risks to assess the total health risk of contaminated sites (see Chapter 8 for an example of how the risk assessment of contaminated sites can operate).

Assumptions about exposure are the critical issue in most risk assessments of contaminated sites. The basic definition of health risk is exposure times toxicity. There is usually considerable consensus on the toxicity of the toxic substances that contaminate a site; it is the estimates of exposure that are difficult. There is a lot we do not know about environmental pathways to exposure, which is a difficult problem because we must make assumptions about how people will use the site in the future in order to determine future exposure. Over time, people may use a specific parcel of land for different purposes. For example, developers used the contaminated Love Canal site for private homes and a school. Many analysts criticize the Superfund program because it assumes that after remediation, people will use the site for residences. As a result, Superfund concludes that the environmental cleanup should make the site sufficiently safe to develop an on-site residential drinking water supply (Karlin, 1994).

HEALTH TRENDS

Some public health trends lend circumstantial support to the concern that toxic substances from contaminated sites and other sources are having a substantial effect on human health. Although the health effects of toxic substances are controversial, there is broad agreement that they are adding to the disease burden in a significant way (U.S. Department of Health and Human Services, 1980).

In the past, environmental health was synonymous with the control of infectious disease agents in food and water. The most economically advanced countries have largely controlled infectious diseases, except for acquired immunodeficiency syndrome (AIDS). Today, chronic diseases such as cardiovascular disorders and cancer are increasingly of primary importance in the economically developed countries and are becoming more important in the developing countries.

The trend in public health in the first half of the twentieth century in the developed countries was a dramatic increase in human life expectancy as infectious and parasitic disease rates dropped sharply. This trend ceased about the middle of the century. Since about 1950, chronic disease rates have increased as the population began to live long enough to die from heart disease, stroke, and cancer. At the middle of the century, industrial societies began releasing large volumes of toxic substances into the environment and the consequent pollution has created the present-day contamination and environmental cleanup problem.

Starting in the mid-1970s, there was wide speculation that carcinogens released into the environment were causing the increase in cancer mortality and the slowing of the rate of increase in the human life span. Many people view the explosive growth of the chemical industry in the post-World War II period as a potential explanation for the change in health trends. During the early post war period, the clear benefits of new chemical products were obvious, but few people recognized the health risks of the new compounds. The new pesticides and herbicides were tremendously effective, and the many new synthetic materials were less expensive than their natural alternatives. They allowed the introduction of a fantastic variety of new and useful products. Of the myriad chemical compounds to which people are now exposed, a U.S. study concluded that almost none has undergone sufficient testing to determine potential toxicity (U.S. National Research Council, 1984). Industry uses these compounds widely in commerce, small numbers of food additives, pesticides, and drugs.

The potential link between environmental toxins, such as those found on contaminated sites, and cancer is the main cause for concern about health trends. Cancer is the second leading cause of death in the United States and other economically developed countries, exceeded only by heart disease. Among the three leading causes of death, only cancer has shown a substantial increase in mortality rates in this century. Some research results suggest that there is a direct link between environmental pollution and cancer. Page *et al.* (1976) suggested that toxic microcontaminants in water supplies caused elevated cancer mortality in certain locations. Pollack and Horm (1980) reported data that suggested that trends in cancer rates had changed and started in increase around the middle of this century. Greenberg (1983) found remarkable changes in cancer rates since 1950 with marked differences by sex, race, and type of cancer in the United States and in other economically developed countries.

Analysis of trends in cancer rates in the past few decades provides additional evidence suggesting that toxic substances may be causing some of the cancer. There has been a dramatic convergence of different rates of disease among cities, suburbs, and rural areas in the United States (Greenberg, 1987). These changes correspond to industrial location and population changes between regions and within metropolitan areas.

CONCLUSIONS

All of the foregoing health studies suggest a possible link between toxic substances in the environment and adverse public health effects, but they do not prove a link. They do raise the perception of risk in both the scientific community and the general public. These and other studies suggest the possibility that toxic chemicals from contaminated sites may sicken future humanity with widespread cancer and genetic mutation. It is not likely that we will have definitive health studies of the risks posed by toxic sites for many years.

On the other hand, it is possible that our present assessment of public health greatly exaggerates the risk caused by contaminated sites. Some research suggests that these sites pose a lesser risk to human health than other environmental problems that receive less attention. In the interim, while we wait for more conclusive health studies, countries must devise and enforce policies to address contamination and environmental cleanup under conditions of great uncertainty.

Countries clearly need policies that will prevent additional toxic contamination and encourage environmental cleanups. The relative importance of the problem and how much of a country's scarce resources should be allocated to this problem are unknown because of insufficient knowledge of the potential human health effects. In any case, it is advisable to make policies as efficient and cost-effective as possible for public health as well as social and economic reasons.

References

Anderson, M. P. 1987. Hydrogeologic framework for groundwater protection, pp. 1–27. In G. W. Page (ed.), *Planning for Groundwater Protection.* New York: Academic Press.

Anderson, R. F. 1987. Solid waste and health, pp. 173–204. In M. R. Greenberg (ed.), *Public Health and the Environment: The United States Experience.* New York: Guilford Press.

Baum, A., R. Fleming, and J. E. Singer. 1983. Stress at Three Mile Island: Applying psychological impact analysis. In L. Brickman (ed.), *Applied Social Psychology Annual.* Beverly Hills, CA: Sage.

Calkins, B. M. 1987. Life-style and chronic disease in western society, pp. 25–75. In M. R. Greenberg (ed.), *Public Health and the Environment: The United States Experience.* New York: Guilford Press.

Clean Sites, Inc. 1990. *Improving Remedy Selection: An Explicit and Interactive Process for the Superfund Program.* Alexandria: VA: Clean Sites, Inc.

Dienemann, E. A., R. C., Ahlert, and M. R. Greenberg. 1991. Remediation of the

Lipari landfill, America's #1 ranked Superfund site. *Impact Assess. Bull.* **9,** 13–30.

Elliot, M. L. 1984. Improving community acceptance of hazardous waste facilities through alternative systems for mitigating and managing risk. *Hazard. Waste* **1**(3), 397–410.

Federal Emergency Management Agency, U.S. Environmental Protection Agency, and U.S. Department of Transportation. 1990. *Hazardous Materials: A Citizen's Orientation* (Reference No. HS-5). Washington, D.C.

Florida Department of Natural Resources Protection. 1994. *Bi-annual Inventory Report of Contaminated Locations in Broward County, Florida.* Fort Lauderdale, FL: Broward County Department of Natural Resources Protection, 97 pp.

Gibbs, L. 1982. *Love Canal: My Story* (as told to Murray Levine). Albany: State University of New York Press.

Greenberg, M. R. 1983. *Urbanization and Cancer Mortality.* New York: Oxford University Press.

Greenberg, M. R. 1987. Health and risk in urban-industrial society, pp. 3–24. In M. R. Greenberg (ed.), *Public Health and the Environment: The United States Experience.* New York: Guilford Press.

Greenberg, M. R., and R. F. Anderson. 1984. *Hazardous Waste Sites: The Credibility Gap.* Piscataway, NJ: Center for Urban Policy Research.

Greenberg, M. R., and G. W. Page. 1981. Planning with great uncertainty: A review and case study of the safe drinking water controversy. *Socio-Econ. Plann. Sci.* **15,** 65–74.

Greenberg, M., and D. Wartenberg. 1992. Communicating to an alarmed community about cancer clusters: A fifty state survey. *J. Commun. Health* **16,** 71–82.

Guerrero, P. 1995. *Superfund: Risk Assessment Assumptions and Issues* (GAO/T-RCED-95-206). Washington, DC: U.S. General Accounting Office.

Hirshorn, J. S. 1984. Siting hazardous waste facilities. *Hazard. Waste* **1**(1), 423–429.

Karlin, A. S. 1994. How long is clean? The temporal dimension to protecting human health under Superfund. *Nat. Resourc. Environ.* Summer.

Myers, S., J. Manno, and K. McDade. 1995. Great Lakes human health effects research in Canada and the United States: An overview of priorities and issues. *Great Lakes Res. Rev.* **1**(2), 13–23.

Office of Technology Assessment, U.S. Congress. 1984. *Protecting the Nation's Groundwater from Contamination,* 2 vols. Washington: DC: U.S. Government Printing Office.

Page, G. W. 1981. Comparison of groundwater and surface water for patterns and levels of contamination by toxic substances. *Environ. Sci. Technol.* **15,** 1475–1481.

Page, G. W. 1987a. Perth Amboy, New Jersey, case study, pp. 289–298. In G. W. Page (ed.), *Planning for Groundwater Protection.* New York: Academic Press.

Page, G. W. (ed.). 1987b. *Planning for Groundwater Protection.* New York: Academic Press.

Page, G. W. 1987c. Drinking water and health, pp. 69–87. In G. W. Page (ed.), *Planning for Groundwater Protection.* New York: Academic Press.

Page, G. W., and H. Rabinowitz. 1993. Groundwater contamination: Its effects on property values and cities. *J. Am. Plann. Assoc.* **59**(4), 471–479.

Page, T., R. Harris, and S. Epstein. 1976. Drinking water and cancer mortality in Louisiana. *Science* **193,** 55–57.

Pigen, B. J. 1984. Methods for assessing health. In M. Harthill (ed.), *Hazardous Waste Management.* Boulder, CO: Westview Press.

Pollack, E., and J. Horm. 1980. *Trends in Cancer Incidence and Mortality in the United States:* 1950–1977. Washington, DC: U.S. Government Printing Office.

Ricci, P. A., and L. S. Molton. 1985. Regulating cancer risks. *Environ. Sci. Technol.* **19,** 473–479.

Slovic, P., B. Fischoff, and S. Lichtenstein. 1979. Rating the risks. *Environment* **21,** 14–36.

U.S. Congress. 1989. Office of Technology Assessment, *Coming Clean: Superfund's Problems Can Be Solved. . .* (OTA-ITE-433). Washington, DC: U.S. Government Printing Office.

U.S. Department of Health and Human Services. 1980. *Health Effects of Toxic Pollutants* (Report prepared for the U.S. Senate by the Surgeon General, Serial No. 96-15). Washington, DC: U.S. Government Printing Office.

U.S. Environmental Protection Agency. 1977. *State Decision-Maker's Guide for Hazardous Waste Management* (SW-612). Washington, DC: U.S. Government Printing Office.

U.S. Environmental Protection Agency. 1987. *A Comparative Assessment of Environmental Problems: Overview Report,* Vol. 1. Office of Policy, Planning and Evaluation.

U.S. General Accounting Office. 1991. *Superfund: Public Health Assessments Incomplete and of Questionable Value* (GAO/RCED-91-178). Washington, DC: U.S. Government Printing Office.

U.S. National Academy of Sciences. 1977. *Drinking Water and Health,* Vol. 1. Washington, DC: National Academy Press.

U.S. National Academy of Sciences. 1980a. *Drinking Water and Health,* Vol. 2. Washington, DC: National Academy Press.

U.S. National Academy of Sciences. 1980b. *Drinking Water and Health,* Vol. 3. Washington, DC: National Academy Press.

U.S. National Research Council. 1984. *Nation's Health.* Washington, DC: U.S. Government Printing Office.

U.S. National Research Council. 1991. *Environmental Epidemiology: Vol. 1: Public Health and Hazardous Waste.* Washington, DC: National Academy Press.

U.S. National Research Council. 1993. *Issues in Risk Assessment.* Washington, DC: National Academy Press.

Wallach, R., and R. Shabtai. 1993. Surface runoff contamination from chemicals initially incorporated below the soil surface. *Water Resourc. Res.* **29**(3), 697–704.

Wartenberg, D., and M. Greenberg. 1992. Methodological problems in investigating disease clusters. *Sci. Total Environ.* **127,** 173–185.

Williard, D. E., and M. M. Swenson. 1984. Why not in your backyard? Scientific Data and nonrational decisions about risk. *Environ. Manage.* **8**(2), 93–99.

CHAPTER 4

Policies to Prevent Additional Contaminated Sites

INTRODUCTION

The prevention of additional contaminated sites is one of the three policy approaches necessary to solve the contaminated sites and environmental cleanup problem. The other two approaches are: taking immediate action to protect the health of people at risk because of proximity to contaminated sites, and cleaning up contaminated sites to make them environmentally and economically productive. Despite progress on the latter two approaches, countries cannot solve their contamination problem unless they develop, enact, and implement effective policies to prevent new contaminated sites.

Because private markets provide little or no incentive to prevent pollution, governments must take action to prevent contaminated sites. Good environmental policies complement economic development, despite common industry claims that they hinder economic growth and international competitiveness (Schmidheiny, 1992). According to a World Bank report, "the evidence indicates that the gains from protecting the environment are often high and that the costs in foregone income are modest if appropriate policies are adopted" (World Bank, 1992, p. 1). Our emphasis will be on "appropriate" policies.

There are benefits from environmental cleanups. The most direct benefits are from reduced negative human health effects, risks, and public fear of contaminated sites. There are also some economic benefits, such as the jobs and wealth creation resulting from large expenditures on environmental cleanups. In addition, cleanups allow the reuse or redevelopment of sites, thus putting them back into productive use and stopping or reversing the processes of urban decay to which contaminated sites contribute.

There are however potential tradeoffs between environmental protection and income and between present benefits and benefits to future generations. The first tradeoff has two aspects: (1) the hundreds of billions of dollars likely to be spent on environmental cleanups of contaminated sites could alternatively be used to promote and assist economic development that would more directly produce jobs and wealth, and (2) when a country takes billions of dollars out of its economy to pay for environmental cleanup, this may cause its goods and services to become more expensive, which makes them less competitive on world markets, and thus has the potential to lower the country's future wealth compared to what it would be without high expenditures for environmental cleanup.

The second tradeoff, between present benefits and benefits to future generations, concerns our obligations for stewardship of the environment. If we spend no money for environmental cleanup, then we will have more wealth to spend on consumption or other uses during our lifetimes. This strategy will pass on the problems of environmental contamination to future generations who will inherit both the contamination and the huge cost of environmental cleanup. The intergenerational inequity of such a scenario is obvious. Many people believe that we have a moral obligation to pass on to future generations an environment that is as healthy as the one we inherited.

Policy makers should strive for sustainable development when they create policies to prevent the creation of additional contaminated land. Creating wealth for the present generation by ignoring the real costs of disposing toxic chemicals will force future generations to bear the potential health risks as well. Furthermore, the costs to remediate contaminated sites are much greater than the costs of appropriate disposal of toxic substances.

Knowledge of how to make effective environmental policy is still evolving. It is only in the past 20 years that countries have made a major effort to control hazardous pollution and protect the natural environment from toxic substances. The policy approaches of different countries have achieved various degrees of success. Some countries, for instance, The Netherlands, have achieved considerable success in negotiated agreements with industries to reduce pollution (Marshall, 1993). Nonetheless, though cooperation is useful, policies must require compliance. Effective approaches generally require a comprehensive and complementary package of policies; a single good policy is generally insufficient.

This review of policy approaches focuses on two main categories: command and control policies and economic incentive (market-based) policies. We then discuss the politics of enacting and implementing effective environmental policy and the role of public participation in the environmental policy-making process. The chapter concludes with a case study of the policy approaches used in the United States to prevent the creation of contaminated sites.

COMMAND AND CONTROL POLICIES

Command and control approaches are the most common means of regulating the use and disposal of hazardous materials and wastes. They require considerable government intervention to determine what controls are best, to implement the system of controls, and to monitor the activities of numerous potential polluters. Critics complain that command and control policies have high costs, are not efficient, consume large amounts of scarce administrative capacity, and stifle innovation. Despite these criticisms, they are the only policy instruments that have achieved significant success in preventing the creation of additional contaminated sites. Other approaches may be at least as successful, but countries have not yet used them sufficiently to evaluate their effectiveness.

Command and control policies usually include restrictions that specify a maximum allowable quantity of pollution. This measure is useful with toxic pollutants because there is often a human-health-based standard that specifies an acceptable health risk. Using health risk models, countries can establish a quantitative health-based standard for the amount of any substance in soil or groundwater. Environmental policy makers then enact a command and control regulatory standard for pollution so that firms do not pollute beyond the level that exceeds this threshold health impact.

Such approaches can be highly restrictive. In some industries when a firm is constructing a new factory, the U.S. EPA provides "new source performance standards" that specify in great detail the technologies that EPA will accept.

Command and control policy may vary from country to country. Many local circumstances determine which policies will be most effective and the specifics of those policies. Experience suggests that the following three factors are especially important.

1. Environmental standards must be realistic and enforceable. This applies to all countries, but is especially important in developing countries that may consider adopting standards used by Organization for Economic

Cooperation and Development (OECD) countries. Unrealistically high standards can result in ineffective implementation, wasted resources, corruption, and loss of credibility for all environmental policies (World Bank, 1992, p. 13).

2. Environmental policies must be consistent with other government policies. Policies to prevent pollution will not be effective if other government regulations promote economic output at any cost (Georgieva, 1992).

3. Most countries will need a combination of policies to prevent contaminated sites. Many firms, organizations, and individuals, undertaking diverse activities, release toxic pollution. All pollution sources and activities must be addressed to achieve effective regulation.

Command and control policies commonly include the following approaches: premarketing product review, product standards, the cradle-to-grave control system, technical and operating standards, permits and licenses, and land use controls (Table 4.1).

TABLE 4.1

COMMAND AND CONTROL POLICIES TO PREVENT CONTAMINATION[a]

Instruments	Advantages	Disadvantages
Premarketing product review	Opportunity to ban or restrict dangerous products before sales	Difficult to operationalize what constitutes dangerous; high testing costs; difficult to obtain proprietary information
Product standards	Great control over regulated products	High implementation and monitoring costs
Cradle-to-grave system	Prevents illegal disposal; audit trail to establish liability	High implementation and monitoring costs
Technical and operating standards	Require safe handling of toxic materials and wastes; bans appropriate uses	High implementation and operating costs
Permits and licenses	Compliance before facility operation; facilitates enforcement of effluent and emissions standards; government control to suspend	High monitoring and enforcement costs
Land use controls	Prevents inappropriate siting of pollution sources	Vulnerable to local political and economic pressure

[a]Modified from Bernstein (1993).

Governments use a product review process to evaluate new products, before they reach markets, to help control toxic pollution. These policies require manufacturers, especially chemical manufacturers, to provide government with product information about toxic ingredients and potential health hazards of new products. Governments can then ban new chemicals or products that they conclude pose an "imminent hazard" to human health. Three examples of such policies are the United States' 1976 Toxic Substances Control Act, Japan's Law concerning Examination and Regulation of Manufacture of Chemical Substances (LERMCS) of 1973 and 1986 (Nakanishi, 1994), and the OECD Minimum Premarketing set of Data (MPD). The potential problems with this approach are determining what testing standards to use, what constitutes an "imminent hazard," and overcoming companies' claims of protecting proprietary information.

Product standards are used to control the entire life cycle of selected products. Officials assign standards to products that contain hazardous materials that they judge to represent a threat to the environment or public health. These standards are then used to ban, control, or restrict such products. Standards can also control the manufacture, sale, import, export, or disposal of such products. Some toxic and persistent pesticides are good examples of the product standard approach in the United States.

The cradle-to-grave control system regulates toxic substances via a comprehensive set of standards, regulations, and requirements from the point of product generation to the final disposal site. These controls apply to hazardous waste generators and transporters, as well as those who store, treat, or dispose of hazardous wastes (Turner, 1992). Generators must determine whether substances are hazardous based on qualities such as ignitability, corrosivity, reactivity, and toxicity. They must also obtain a firm identification number and a permit for the generating facility. During shipment, this permit must accompany the materials in appropriate shipping containers and with a shipping manifest that tracks the material's movements from the time it leaves the generating facility until final disposal. Governments use this policy approach to minimize illegal disposal of hazardous wastes, sometimes referred to as "midnight dumping."

Governments can also use technical and operating standards to control many aspects of hazardous waste disposal. Hazardous waste transporters must meet standards for labeling, packaging, placarding, and transport of the waste, and they are responsible for tracking the movement. Transporters must also report any discharges or spills during transit and are responsible for their cleanup. Technical standards specify the design, construction, maintenance techniques, and pollution control technologies that hazardous waste storage, treatment, and disposal facilities can use. Standards may include limits on air emissions or even bans on certain forms of disposal, most often land disposal. For example, the United States bans waste or used oil containing dioxin and other specified contaminants from use as dust-control agents on dirt roads.

Another approach is to use permits and licenses to ensure the safe operation of hazardous waste treatment, storage, and disposal facilities. Such permits ensure that facilities meet established standards. For example, the U.S. Resource Conservation and Recovery Act of 1976 and its amendments set standards for facility design, engineering, operating procedures to prevent contamination, groundwater protection measures, contingency plans for emergencies, waste analysis procedures, inspection schedules, and closure and postclosure plans. The license may also specify that government environmental agency inspectors can enter the facility to conduct inspections, sample wastes, and examine records.

Local planning officials use land use controls such as zoning to ensure that the locations of waste disposal facilities do not pose a direct threat to other activities. Officials can use land use controls to site facilities, particularly to restrict siting to areas with appropriate soils and geological conditions and prevent siting near aquifer recharge areas, well fields, residential areas, or sensitive land uses such as schools or hospitals.

ECONOMIC INCENTIVE POLICIES

In recent years, several countries have begun experiments with economic incentive policies to help solve environmental problems. From the perspective of economic theory, these incentive approaches have the potential to be more “efficient,” to achieve greater environmental protection at less cost and with fewer distortions to the overall economy. These policies allow polluters to choose their method of pollution control, thus promoting the most cost-effective approach. Economic incentives also reduce the cost of compliance, are often administratively less complex than command and control regulations, and encourage innovation in waste-minimizing technologies.

In theory, economic incentive policies use market mechanisms, thereby facilitating deregulation and a reduction in government involvement. They are especially valuable in developing countries because they do not demand such a large government administrative capacity, which is often scarce. However, in all countries that have used economic incentive policies, these approaches have not eliminated the need for government involvement and command and control environmental policies (Bernstein, 1993). Incentive approaches still require government involvement with standards, environmental monitoring, and enforcement. Implementation difficulties have limited the effectiveness of incentive approaches, so they should be viewed as effective supplements in a comprehensive package of policies designed to prevent contaminated sites.

Several types of market-based incentives are commonly used: disposal charges, product charges, administrative charges, liability insurance, subsidies, and enforcement incentives (Table 4.2).

Disposal charges are direct taxes or fees on hazardous wastes. Government can collect the disposal charge at either the point of generation or

TABLE 4.2

ECONOMIC INCENTIVE POLICIES TO PREVENT CONTAMINATION[a]

Instruments	Advantages	Disadvantages
Disposal charges	Raise revenue; encourage polluters to reduce discharges; encourage innovation in control technology; promote cost savings	Encourage illegal disposal; encourage export of hazardous wastes; high implementation cost
Product charges	Raise revenue; promote use of safe products; incentive to find less toxic substitutes	Require close substitute products or inputs
Administrative charges	Raise revenue; some control over facilities; encourage use of safe products	Limited applications
Liability insurance	Incentive to control or clean up pollution	Complex implementation; high monitoring cost
Subsidies	Incentives to control pollution; incentive to manage wastes; promote innovative technologies; low monitoring costs	Impose costs on tax payers rather than polluter; perpetuate polluting firms
Waste reuse markets	Encourage recycling and waste minimization; require little government involvement	Sophisticated testing and waste handling required
Noncompliance fees	Encourage compliance; high administrative costs	Difficulty setting penalties at the appropriate level
Performance bonds	Ensure environmental cleanup	Limited applications
Liability awards	Encourage firms not to pollute	Costly litigation; long delays

[a]Modified from Bernstein (1993).

the point of disposal. Economists refer to disposal charges as waste-end-taxes. These charges provide an economic incentive to firms and individuals to eliminate or minimize the hazardous wastes they send to disposal. This incentive promotes input substitution and process changes to achieve waste reduction and encourages sending hazardous wastes to recycling or incineration rather than disposal sites, which are usually landfills. Policy makers must ensure that disposal charges do not promote illegal dumping of hazardous wastes or their export to developing countries that may lack policies to prevent contaminated sites. Governments must coordinate disposal charges with policies that promote the development of recycling and incineration facilities, discourage midnight dumping, and control the export of hazardous wastes. The administrative costs of implementing a disposal charge on hazardous wastes may be high. Both the United States and The Netherlands have used disposal charges on hazardous wastes, but analysts have judged neither attempt to be successful (Bernstein, 1993, p. 59).

Product charges provide an economic incentive to firms and individuals to eliminate or minimize the production of specified products that become hazardous wastes. Oils and other lubricants are an example. Used oil often contains toxic compounds that may create contaminated sites if not disposed of safely. In theory, a product charge on those lubricants that become hazardous wastes will produce a market incentive to minimize such lubricant uses and promote innovations to find nonhazardous waste-producing lubricants. Austria and Sweden use product charges for pesticides to provide an incentive to the agricultural sector not to overuse pesticides (World Bank, 1992, p. 76). France uses a product charge on lubricants to fund infrastructure to collect, store, and properly dispose of used oil. However, this product charge is too low to have incentive effects (Bernstein, 1993, p. 59).

Governments may collect administrative charges for services such as chemical registration or licensing of products, which in turn finance the service provided. Norway uses such fees to finance the licensing of chemical products (Bernstein, 1993, p. 12). Belgium has a mandatory registration fee for imported and exported waste (Bernstein, 1993, p. 13; Organization for Economic Cooperation and Development, 1989). Administrative charges may be based on the relative hazard of the product to provide an incentive for the use of less hazardous products.

Liability insurance acts as an incentive to reduce the use and disposal of hazardous materials and wastes because the cost of the insurance increases with the risk of leak, spill, or improper disposal. In countries that use environmental liability to finance cleanup, liability insurance costs can be high. For some firms with high risk of causing contaminated sites, insurance firms may refuse to provide liability insurance.

Governments can also use subsidies as an incentive to help prevent contaminated sites. Subsidies are government financial incentives to achieve a specific goal, for instance, pollution control, recycling, or resource recovery. They include grants, low-interest loans, and tax incentives. Policy can direct subsidies to specified products that do not produce hazardous wastes and that would compete with hazardous waste-producing products, or it can direct subsidies to specific technologies, such as waste minimization, pollution abatement, or remediation. Subsidies can also support research or training programs. In general, subsidies have a high potential to create negative distortions in an economy that are not always easy to predict (Kosmo, 1989).

Waste reuse markets provide incentives to recycle hazardous wastes. They match waste products generated by firms to the input needs of other firms and establish the price. Where markets can make matches, both firms find this approach to be cost-effective. In some locations, these markets work well for a small range of hazardous waste products.

Enforcement incentives can be an especially powerful approach to preventing contaminated land. They generate economic incentives by enforcing command and control regulations, often through noncompliance fees or fines, performance bonds, and liability awards. These "costs" are the incentives to minimize future pollution. Policies that allocate the cost of remediating contaminated sites to the firms that caused them provide strong economic incentives against future toxic pollution. This is a major component of the U.S. Superfund program. The issue of environmental liability is a major topic of subsequent chapters.

THE POLICY-MAKING PROCESS

The policy-making process itself has a strong effect on efforts to prevent contaminated sites. Each country has unique political factors and some of these factors vary over time. An understanding of the policy-making process and how political factors influence policies is important in evaluating the policies of different countries.

There are many variations in the policy-making process, but the basic six stages are presented in Fig. 4.1. The first step is problem formation, also known as agenda setting. In this stage, the problem gets recognized. In Chapter 1, the U.S. case study considered the role of the mass media in publicizing notorious contaminated sites, such as Love Canal. The extensive media coverage of a small number of contaminated sites galvanized public opinion and helped formulate the more broadly recognized

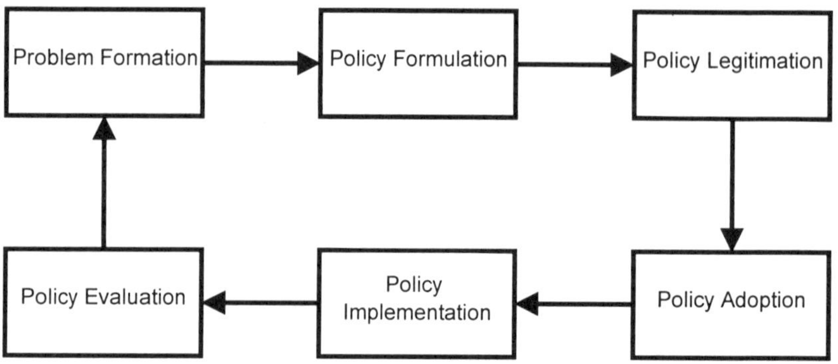

FIGURE 4.1 The policy-making process.

problem of contaminated sites as sufficiently serious to warrant government action.

The second stage of the policy-making process is policy formulation. During this stage, people suggest and evaluate alternative ways of addressing the problem. These proposed approaches can come from a wide variety of sources within and outside the government.

The policy legitimization stage refers to all the activities that take place to gain political support for the different proposals. Many individuals and organizations are active in this stage, which often extends throughout the duration of the policy-making process. Those favoring different approaches may include government officials of several agencies, industries that produce or use hazardous materials, the waste disposal industry, the waste transporting industry, environmental interest groups, the real estate industry, the insurance and banking industries, and others.

The fourth stage of the process is policy adoption. This is the formal selection of one or some combination of proposed actions as the official policy. Each country requires official action, but that action varies according to political system. Policy adoption can take place at the international, national, regional, or local government levels.

The policy implementation stage is critical to effective policy. It refers to the administrative actions that render policies operative after governments officially adopt them. As a generalization, command and control environmental policies require considerable administrative action during the implementation stage, whereas economic incentive policies require less. In some countries, the legislative branch of government that enacts legislation and the executive branch that implements legislation have different views of the environmental problems. Such differences of opinion seriously affected the implementation of policies to prevent contaminated sites during the Reagan administration in the United States (Portney, 1992, p. 144).

The final stage of the policy-making process is policy evaluation, during which analysts evaluate the effectiveness of policy. The results of this evaluation should then be used as feedback into the problem formulation stage, thus forming a continuous policy-making process that can determine whether the problem is getting worse or better. Officials may recognize unintended consequences of policy at this stage, and policy makers may modify the policy, or discrete components of the policy, in response to the feedback from the policy evaluation stage. For example, in the United States, the Hazardous and Solid Waste Amendments Act of 1984 represents a major redirection of policy from the 1976 Resource Conservation and Recovery Act. Using policy evaluation feedback, the 1984 act greatly increased the regulation of hazardous wastes to prevent contaminated sites.

Public Participation in the Policy-Making Process

Public participation in the policy-making process may be a major obstacle or a major asset in developing policy to prevent the creation of new contaminated sites. Involving the public in environmental policy making can be expensive, may produce stalemates when local activists refuse to compromise, and may reinforce local elites who have an agenda that is different from national environmental goals. Enforcing policies that the public does not accept is rarely successful. For example, public fear of hazardous waste processing and disposal facilities may obstruct the siting of such facilities and thereby contravene policy to prevent contaminated sites.

When the process encourages public participation from the beginning of the policy-making process, it can make a significant positive contribution. In some cases, public participation is essential because the government cannot successfully implement programs if the local population does not accept them. There are three primary benefits from public participation in the environmental policy-making process (World Bank, 1992, p. 93):

1. Public participation gives environmental planners and policy makers a better understanding of local values, knowledge, and experience. Community support of policy is unlikely unless that policy reflects local beliefs, values, and ideology.

2. Public participation helps win community backing for project objectives and community help with local implementation. Community support for projects often leads to successful projects. Studies by the World Bank

and the U.S. Agency for International Development report strong correlation between levels of public participation and project success (World Bank, 1992, p. 94).

3. Public participation can help resolve conflicts. Involving community activists in the policy-making process often leads to participants who are realistic about what public policy can accomplish and are willing to compromise on issues to produce effective policy (Beatley *et al.*, 1994).

POLICY APPROACHES IN THE UNITED STATES TO PREVENT ADDITIONAL CONTAMINATED SITES

In the 1970s, contamination and potential public health problems gained wide notoriety and media coverage in the United States. In response to public pressure, Congress passed legislation to regulate the use and disposal of toxic and hazardous materials in an effort to prevent the contamination of additional sites (Table 4.3).

Policy in the United States has evolved quickly over the past two decades and continues to change. Prior to environmental policy initiatives to control the causes of contaminated land, industry carelessly disposed of hazardous wastes to minimize their costs. These disposal practices were legal until the passage of the Resource Conservation and Recovery Act (RCRA) of 1976 and the Toxic Substances Control Act (TSCA) of 1984. These acts had three major components: minimize the dumping of hazardous wastes, treat hazardous wastes, and source reduction.

TABLE 4.3

OVERVIEW OF MAJOR U.S. LEGISLATION AFFECTING INDUSTRIAL WASTE DISPOSAL

Act	Year	Description
Resource Recovery Act (RRA)	1970	Introduced federal role, research, and demonstration projects
Resource Conservation and Recovery Act (RCRA)	1976	Control of hazardous waste
Comprehensive Environmental Response, Compensation, and Liability Act (CERCLA)	1980	"Superfund" program to clean up contamination
RCRA Amendments	1984	Better implementation and control of underground leaks
Toxic Substances Control Act (TSCA)	1984	Control new substances before they enter the marketplace
Superfund Amendments and Reauthorization Act (SARA)	1986	Modified and strengthened CERCLA

Minimize Dumping of Hazardous Wastes

The first policy approach was to control the movements of hazardous wastes and to minimize careless dumping of hazardous wastes on the land. In 1984, TSCA established a registration system for toxic substances before their manufacture. In 1976, RCRA regulated hazardous waste generators, transporters, and disposal facilities. It established a cradle-to-grave manifest system designed to track hazardous wastes and ensure their disposal in licensed facilities. RCRA also required states to develop solid waste management plans.

Treating Hazardous Wastes

The second policy approach, the treatment of hazardous wastes, quickly evolved with the recognition of the inherent weaknesses of land disposal facilities. The "minimize dumping" approach resulted in huge volumes of hazardous wastes being sent to landfills, which was obviously not a satisfactory solution because all landfills will eventually leak. The second policy approach emphasized the treatment of hazardous wastes. Proven treatment technologies were well known and readily available to greatly reduce the hazardous waste stream.

Technologies to remove metals from liquids, to neutralize acids, to recycle oils and solvents, and to incinerate (via thermal destruction) some chemicals were the most common. It is estimated that industries can effectively treat 80–90% of all hazardous wastes in the United States using modern variants of 1930s-era wastewater treatment technologies (Mazmanian and Morell, 1992). While land disposal of hazardous waste was the preferred low-cost disposal method, treatment technologies were still not cost-effective and industries seldom used them. As environmental policy to prevent contaminated land increased the cost of land disposal through regulations, treatment became more widely used.

The 1984 RCRA Amendments greatly increased the cost of land disposal by setting technical standards for land disposal facilities that included the requirement of double liners, leachate control systems, and groundwater monitoring systems. These amendments also restricted the land disposal of solvents. This legislation promoted the treatment of hazardous wastes by making land disposal more expensive.

The treatment approach has achieved considerable success, but by itself cannot prevent additional land from becoming contaminated. There are many examples of firms that have achieved remarkable records of reducing their hazardous waste through treatment. Large firms with expertise, capital, and the potential for large savings were the first to invest in hazardous waste treatment technologies. However, most treatment technologies reduce the volume of hazardous waste but create concentrated hazardous residues such as ash or sludge. Industries usually send these residual wastes to land disposal because that remains the least-cost disposal option. Small firms, on the other hand, had to send their hazardous wastes to treatment facilities because they lacked the economies of scale to construct their own on-site facilities.

The Western European countries have also made progress with hazardous waste treatment facilities. The European Union issued a Directive on Landfill Waste in 1991 to help prevent toxic substances contamination. This directive established guidelines for locating, designing, operating, closing, and long-term monitoring of landfill sites. It called for eliminating the practice of "co-disposal," in which landfills mix municipal solid waste with industrial hazardous wastes. This co-disposal portion of the directive created political discord among the member-states of the European Union that has stalled enacting the directive as legislation. The expansion of the European Union to 15 member-states in January of 1995 changed the political dynamics, and this directive now is likely to have sufficient political support. European countries, for example, Sweden, have had success in siting and constructing centralized hazardous waste treatment facilities (Lidskog, 1993) European countries have also constructed treatment facilities in dispersed locations to treat hazardous wastes near the industries that generated them.

In the United States, firms constructed relatively few treatment facilities for the volume of hazardous wastes generated. The greatest obstacle was public opposition to facility siting. This opposition was especially intense because local populations saw themselves as victims of corporate and government manipulation. They condemned government policy that pressured them to accept hazardous wastes generated in other locations. In the United States, town, cities, and counties make land use decisions about zoning and building permits to construct facilities. In most cases, local governments refused to allow the construction of hazardous waste treatment facilities (Anderson and Greenberg, 1982). The argument of "not in my backyard" (NIMBY) quickly gained momentum throughout the country. As opposition to the siting of hazardous waste facilities collapsed, the term became "not in anybody's backyard" (Heiman, 1990). Liability problems also deterred the construction of hazardous waste facilities. The liability issue made closure and postclosure insurance for hazardous waste disposal operations all but impossible to obtain (U.S. General Accounting Office, 1994).

Increasing the cost of land disposal of hazardous waste has both beneficial and negative aspects. In theory, the cost to pollute the land with hazardous wastes should be at least equal to the "damage" to the environment and human health that the pollution causes. Though the costs of this damage are difficult to measure, they are probably much greater than present disposal costs. Certainly the costs to remediate contaminated land far exceed the present cost saved by polluting the land by disposing hazardous wastes on it.

Unfortunately, as the cost of disposal in licensed landfills increases, so does the economic incentive to illegally dispose of hazardous wastes in ways that produce more contaminated sites. Illicit "midnight" dumping of hazardous wastes on land and in water bodies increases as disposal cost rises, as does shipping the wastes to countries that lack effective policies to protect land from contamination. There is some evidence that fraudulent disposal operations are effectively disguising their illegal operations (Rayer, 1995).

Regulating and treating hazardous wastes are pollution control approaches that assume all substances have some acceptable level of discharge. Many argue that industry should not generate, use, or release any amount of persistent toxic substances. This argument is based on the persistence of the substances in the environment and either known effects or the lack of data on the potential effects of these toxic substances. The U.S. TSCA provides the authority to ban, phase out, and restrict and regulate the use of toxic substances. Similar legislation exists in other countries, for example, the LERMCS legislation in Japan (Nakanishi, 1994). The U.S. government has banned few chemicals, and most are not complete bans on the substance. For example, the United States bans polychlorinated biphenyls, although the government still allows their use in some closed systems like transformers. The United States also bans the pesticide DDT, but U.S. firms may still manufacture it for export.

A policy designed to achieve the virtual elimination of persistent toxic substances is an alternative approach to a pollution control policy. The International Joint Commission (IJC) between Canada and the United States adopted a virtual elimination policy to help clean up the Great Lakes. The IJC proposed this policy goal to stop present inputs and prevent future inputs of those substances for which the Great Lakes ecosystem has no assimilative capacity (International Joint Commission, 1991). The IJC policy has three components:

1. Anticipation and prevention, which can be viewed as the zero input route. The intent is to reduce or deny entry of a persistent toxic substance into the ecosystem by disallowing its manufacture, production, or use.
2. Remediation focuses on cleanup after contamination has occurred.
3. Treatment focuses on denying entry of a persistent toxic substance into the ecosystem after that substance has been produced or used (International Joint Commission, 1991, p. 17).

Source Reduction

While many have praised this policy goal, it has had limited impact. The zero input of toxics has been the least successful component of the policy goal. Without a well-conceived implementation plan and the requisite regulatory powers, policy goals have limited utility.

As U.S. policies evolved, the goal gradually shifted to eliminating persistent toxic substances to the extent feasible. Without abandoning support for minimizing dumping and treating hazardous wastes, a third U.S. policy approach of source reduction thus gained support. Source reduction entails changing processes to use less hazardous materials and produce fewer hazardous wastes. As the costs of handling hazardous materials and disposing of hazardous wastes increased, source reduction became cost-effective for many firms. Techniques to achieve source reduction include process change or modification, raw material substitution, product reformulation, screening new chemicals, and bans.

This policy approach also forces industries to make publicly available information on the uses, emissions, leaks, and spills of selected hazardous substances. The Superfund Amendments and Reauthorization Act (SARA) of 1986 Title III, and the Emergency Planning and Community Right-to-Know Act of 1986 require firms to release this information annually. Concern over the Bhopal accident in India in 1984 influenced the effort to include the right-to-know requirements in SARA. Forcing firms to publish information may have an effect on behavior and cause some firms to use less hazardous materials and produce fewer hazardous wastes. This information also allows individuals and nongovernmental organizations to monitor emissions against allowable permit levels.

Some firms were very successful in finding alternative materials and in reducing hazardous wastes. However, only a relatively small number of large and technologically sophisticated firms have achieved success in source reduction. Suitable replacements for many hazardous materials used in industrial processes are not available. When taken to the extreme position of changing processes so that they produce no hazardous wastes, such policy is referred to as a "zero discharge" policy or "sunsetting." Another term for this policy is "virtual elimination," which recognizes that zero discharge may not be technologically possible, but that a policy goal of achieving no discharges that cause a pollution problem is a realistic goal.

CONCLUSIONS

Policy to prevent toxic contamination has evolved rapidly in many countries. Effective policies have drastically reduced illegal or dangerous dumping in favor of controlled land disposal, treatment, and source reduction. The most progressive policies now favor reducing the hazardous waste stream as much as is feasible and then treating what remains.

To some extent, these policy approaches are in conflict. Source reduction decreases the volume of hazardous waste and thereby reduces the economic incentive to construct hazardous waste treatment facilities. In a dynamic policy environment, firms are wary of making large investments in hazardous waste treatment facilities. This reluctance is increased by the legal liability concerns for the owners and operators of treatment facilities (see Chapter 5).

Even with powerful environmental legislation in many countries, new episodes of toxic substances pollution occur that are creating substantial numbers of new contaminated sites. Effective implementation of policies to control hazardous materials and hazardous wastes is more difficult to achieve than enacting the legislation. It is also difficult to overcome public opposition to the siting of hazardous waste treatment facilities, the NIMBY syndrome. Neither private firms nor the government have been able to construct sufficient hazardous waste treatment capacity with the convenient locations that policy makers originally anticipated. Small- and medium-sized firms have achieved little source reduction. For example, some analysts estimated that by the end of the 1980s, U.S. industries had reduced the volume of hazardous wastes by only 10% (Mazmanian and Morell, 1992, p. 18). Progress since then has been marginal.

Pollution with hazardous wastes continues to create new contaminated sites. The 1988 U.S. Toxic Release Inventory reported that 19,000 facilities emitted about 6.2 billion pounds of toxic chemicals to the environment. The Office of Technology Assessment estimated that a startling 95% of total toxic emissions may go unreported (U.S. Government Accounting Office, 1991).

Despite strict regulations, some industrial or commercial sites still experience leaks and spills, and some firms continue illegal dumping. Mining and agriculture also create new contaminated sites. Because of the high costs of compliance and the difficulties of enforcement, policy in most developed countries to prevent the contamination of additional sites has achieved only limited success. Industry in the United States still disposes of the large majority of hazardous wastes in landfills, surface impoundments, and injection wells (Mazmanian and Morell, 1992, p. 14). Because toxic pollution remains a serious problem, environmental cleanup policy using environmental liability for damages and the remediation of contaminated sites plays a potentially significant role in preventing additional contamination. The fear of liability for the high cost of environmental cleanup is a powerful force to prevent further contamination.

References

Anderson, R., and M. Greenberg. 1982. Hazardous waste facility siting. *J. Am. Plann. Assoc.* **48**(2), 204–218.

Beatley, T., D. Brower, and W. Lucy. 1994. Representation in comprehensive planning. *J. Am. Plann. Assoc.* **60**(2), 185–196.

Bernstein, J. D. 1993. *Alternative Approaches to Pollution Control and Waste Management.* Washington, DC: World Bank, in conjunction with the United Nations.

Georgieva, K. 1992. Environmental policy in southeastern Europe: Green desires, gray realities. *Environ. Impact Assess. Rev.* **12,** 239–243.

Heiman, M. 1990. From "Not in My Backyard!" to "Not In Anybody's Backyard!": Grassroots challenge to hazardous waste facility siting. *J. Am. Plann. Assoc.* **56**(3), 359–362.

International Joint Commission. 1991. *Persistent Toxic Substances: Virtually Eliminating Inputs to the Great Lakes.* A Summary Report of the Virtual Elimination Task Force on the International Joint Commission, Detroit, Michigan, April.

Kosmo, M. 1989. *Economic Incentives and Industrial Pollution in Developing Countries.* Environment Department Working Paper No. 1989-2. Washington, DC: World Bank.

Lidskog, P. 1993. Whose environment? Which perspective? A critical approach to hazardous waste management in Sweden. *Environ. Plann. A* **25**(4), 571–588.

Marshall, T. 1993. Regional environmental planning: Progress and possibilities in Western Europe. *Eur. Plann. Stud.* **1**(1), 69–90.

Mazmanian, D., and D. Morell. 1992. *Beyond Superfailure: America's Toxics Policy for the 1990s.* Boulder, CO: Westview Press.

Nakanishi, J. 1994. Regulatory actions for new chemicals under the law (LERMCS) in Japan. *Water Rep.* **4**(5), 1–3.

Organization for Economic Cooperation and Development. 1989. *The Application of Economic Instrument for Environmental Protection,* Environmental Monograph No. 18. Paris.

Portney, K. E. 1992. *Controversial Issues in Environmental Policy.* Newbury Park, CA: Sage.

Rayer, T. 1995. Sham disposal operations on the rise. *Environ. Prot.* **6**(5), 52–54.

Schmidheiny, S. 1992. *Changing Course: A Global Business Perspective on Development and the Environment.* Business Council for Sustainable Development. Cambridge, MA: MIT Press.

Turner, P. L. 1992. Hazwaste transport manifests. *Environ. Prot.* **3**(10), 12–16.

U.S. General Accounting Agency. 1991. *Toxic Chemicals: EPA's Toxic Release Inventory is Useful but Can Be Improved* (GAO/RCED-91-121). Washington, DC: U.S. Government Printing Office.

U.S. General Accounting Office. 1994. *Hazardous Waste: An Update on the Cost and Availability of Pollution Insurance* (GAO/Pemd-94-16). Washington, DC: U.S. Government Printing Office.

World Bank. 1992. *World Development Report: Development and the Environment.* New York: Oxford University Press.

CHAPTER 5

LEGAL ASPECTS OF THE CONTAMINATION AND ENVIRONMENTAL CLEANUP PROBLEM

INTRODUCTION

Experience has demonstrated that countries cannot solve their environmental cleanup problems without government intervention. Effective environmental policies require legal provisions to enable governments to enforce the policies. The choice of legal provisions will in turn critically affect the success of the policies.

Environmental cleanup policy requires legal powers to deter the creation of new contaminated sites, to force the cleanup of dangerous existing sites, and in many countries to provide a means of collecting funds to pay for the environmental cleanup. The extent of the environmental cleanup problem, with the large numbers of contaminated sites, the human health risks, and the threats to the environment that they cause, creates difficult legal problems. Legal approaches to preventing additional contaminated sites were discussed in Chapter 4. This chapter explores the legal dimensions of environmental cleanup policies and focuses on the legal methods to force some parties to pay for the cleanup.

FUNDING APPROACHES

The determination of how to obtain sufficient funds to pay the cost of environmental cleanup is an important issue, and may be the single most critical issue for countries when they develop cleanup policies. It is difficult to overestimate the severity of the problem. For example, the cost of remediating contaminated sites in the United States could range from $400 billion to $1.7 trillion during the next 30 years (Russel *et al.*, 1991).

This section discusses alternative policy options and some of the legal approaches by which governments can obtain cleanup funds. However, it is important to remember that these policies may significantly affect the likelihood that firms will continue to create pollution and thus new contaminated sites. Analysts must evaluate proposed policies to consider how they will both prevent future pollution and encourage cleanup of existing contaminated land.

Liability is the legal term that means legal responsibility, in this context the responsibility to pay for cleanup. If policies are to be effective, they must include mechanisms that force private parties to pay for cleanup or that allow governments to raise sufficient funds for this purpose. In most countries, especially democratic countries, the public must perceive that these policies are "fair." Determining who will pay for the environmental cleanup of contaminated sites can be quite difficult.

There are two principal policy approaches to pay for environmental cleanup: (1) have the firms and individuals who caused the contamination pay for cleanup—the "polluter pays principle"—or (2) have the government pay for cleanup with funds from general revenues. With appropriate laws and implementation, it is possible that either of these approaches can generate the large sums of money necessary and can achieve the public perception that such a policy approach is "fair."

Government Funding of Cleanups

Having the government assume responsibility for environmental cleanup and pay for it with general revenue funds is an efficient approach. It is efficient in the sense that governments can clean up contaminated sites at a lower cost and in less time than if those same governments attempt to force those who caused the contamination to pay for the cleanup. This approach takes less time because it avoids the necessity of identifying the responsible parties, negotiating their respective contributions to the contamination, collecting the funds, and then proceeding with cleanup. Many countries have mechanisms in place whereby the government first cleans up sites and collects the funds afterward for those contaminated sites that represent an immediate and serious health risk (see Chapters 10 and 11 for case studies of such policies).

Paying for environmental cleanup from general revenues is less costly than forcing those who caused the contamination to pay for its cleanup.

The total cost is lower because almost all governments have tax collection systems that produce general revenues with a relatively low expenditure of funds. Because governments can spread the cost of cleanup through value-added taxes, other business taxes, or to all taxpayers in the country by income taxes, each individual's share of the cleanup cost is small.

Even though the use of a country's general revenues to pay for environmental cleanup is efficient, many people do not think it is fair. Despite implementation problems, there is broad political support for the idea that polluters who benefited from causing the contamination should be held responsible. Unfortunately, governments cannot always identify and locate the firms and individuals who caused the contamination, and even when polluters are identified, they are often unable to pay for the cleanup. Yet in a broad economic sense, because entire countries benefited from the increased production and lower consumer prices that resulted from the improper disposal of toxic substances, it may be appropriate for cleanup to be paid from general revenues.

Most countries have enacted environmental cleanup policies that claim adherence to the polluter pays principle, which dominates political debate. However, a careful look at the actual financing of environmental cleanups reveals that funding from government revenues has been a substantial and growing portion of the total expenditures.

Private Party Funding of Cleanups

The polluter pays principle states that the private parties that created a contaminated site must pay for the subsequent environmental cleanup. Such polluters are usually private firms in capitalist economies, but may be government-owned firms or branches of government such as the military. This approach has two advantages: (1) it acts as a strong deterrent to the creation of additional contaminated sites and (2) most people believe that it is "fair" for polluting parties to bear the cost of cleanup. Both this presumed deterrence to new contaminated sites and the broad public support have produced strong political pressures for cleanup legislation based on the polluter pays principle.

For many specific contaminated sites, such a policy can generate sufficient funds and operate efficiently. If an identified firm caused the contaminated site, this policy forces them to pay for the environmental cleanup, which the firm treats as a cost of doing business. Unfortunately, because of the exorbitant cost of cleanup, there are many contaminated sites where the polluter pays principle cannot generate sufficient revenues. The average cleanup cost at U.S. Superfund sites has increased to more than $30 million, which may exceed the net value of some firms. Such liability may drive them into bankruptcy without generating the needed funds to complete the cleanup.

In some cases, governments can no longer find the firms or individuals who caused the contaminated site. Indeed, many sites were created decades ago, and the responsible firms have since moved, merged with other firms, or gone out of business. Furthermore, the shareholders who may have benefited from these actions may no longer be the current shareholders.

Another aspect of this problem is that the release or disposal of toxic substances to the land was once a legal activity. Many firms and individuals claim that it is not fair to make them pay the cost of environmental cleanup for actions that at the time did not violate a statute or regulation. In some cases, these actions had been approved by government-issued permits. Although most people believe it is fair to force parties who break the law and pollute to pay for the cost of cleanup, many people do not think it is fair to force parties to pay for cleanup when they were not guilty of breaking a law.

The polluter pays principle also encounters difficulties when more than one firm or individual caused the contamination at a site. Many of the most dangerous contaminated sites were formerly waste disposal sites or industrial or municipal landfills that accepted waste from many parties. It is obviously a difficult task to assign the correct portion of liability in such a case. To further confound the problem, few records exist of past activities at contaminated sites. It is difficult to determine who disposed of what substances at a site, and especially how much of the toxic contamination each responsible party caused. Sometimes, employees who could help estimate the relative contributions of toxic wastes cannot be located.

The human health and environmental risks of different toxic substances vary, and the cost of remediation also varies widely for these substances. The quantities, toxicity, persistence, and mobility of the disposed substances are often unknown and quite variable. Allocating a fair portion of the environmental cleanup cost to each responsible party, given these difficulties, is nearly impossible. For all of these reasons, when multiple parties are involved, it is difficult to use the polluter pays principle. When it is applied, there are great delays in the cleanup and a significant increase in the cost because of the protracted and expensive efforts to determine each party's liability. Economists refer to these administrative and legal costs as "transaction costs," which are in addition to the cost of the direct remediation work. At some Superfund sites, transaction costs have consumed more than half of the cost of the environmental cleanup (see Chapter 8). High transaction costs make it more difficult to obtain sufficient funds and also undermine political support for cleanup policies.

Thus, policy that relies on the polluter pays principle requires strong administrative and legal powers that are capable of identifying the responsible parties and then forcing them to pay for cleanup.

LEGAL TOOLS

The U.S. Superfund program uses the court system to assign specific liability for the environmental cleanup of contaminated sites. These legal powers maximize the effectiveness with which the polluter pays principle generates funds. The U.S. Comprehensive Environmental Response, Compensation, and Liability Act (CERCLA), popularly known as the Superfund program, casts a wide net in determining who may be responsible for the cost of cleanup.

CERCLA defines four classes of "potentially responsible parties": (1) the current owner or operator of a site, (2) any person who formerly owned or operated the site at the time of disposal of any hazardous substance, (3) any person who arranged for disposal or treatment of hazardous substances at the site, and (4) any transporter of hazardous substances to the site. Court interpretations of these categories make virtually anyone involved with a Superfund site a potentially responsible party, who may be liable for the cost of environmental cleanup.

The Superfund program operates in addition to the common law tort system (toxic tort) that traditionally provides a mechanism for compensating victims of property and health injuries. The toxic tort system developed from English common law. Several countries that are former colonies of the United Kingdom use this system.

CERCLA also defines liability in innovative ways that extend the ability of the Superfund program to obtain the necessary funds: (1) strict liability, (2) retroactive liability, and (3) joint and several liability. These legal tools significantly expand liability beyond its traditional definition. Though use of these tools increases the funds available for cleanup, they create problems in the perception of fairness as well as strong political opposition.

Strict liability extends responsibility for the cost of cleanup beyond the requirement for culpable behavior. Under strict liability, firms and individuals can be liable for cleanup even though they may have broken no law nor been negligent under traditional legal standards (Oswald, 1993).

Retroactive liability extends liability to past actions. In this case, firms and individuals are liable even if the past activities that caused the contaminated sites were legal when they occurred. Many people question the fairness of retroactive liability, and some countries enact cleanup policies with clearly identified dates to establish a point before which governments may not assign liability (see Chapter 10).

Britain's common law system is unclear regarding retroactive liability. There are many examples of courts finding firms and their insurance companies liable for the cost of cleanup of contaminated sites resulting from activities that occurred as long ago as the 1950s (House, 1993). An important precedent is a 1992 decision in which a tannery was held liable for £1 million in damages. The court found the tannery liable for the release of toxic chemicals prior to 1976, at a time when those releases were in compliance with all statutes and regulations and were not legally negligent (House, 1993). For other examples, see the case studies of The Netherlands and the United Kingdom in Chapters 10 and 11, respectively.

In the context of contamination and environmental cleanup policy, joint and several liability can make any one of the potentially responsible parties liable for the entire cleanup cost at a site. Thus joint and several liability provides the government with a powerful tool to collect the funds needed for cleanup, particularly when the government cannot find the parties who contributed most of the toxic substances to the site or when these parties are not able to pay for the cleanup (Wilkerson and Church, 1989). In these situations, the amount that the parties pay for the cleanup may be unrelated to the proportional share of that party's contribution, even when they have minimal connection to the contamination event.

Joint and several liability is an important component of the U.S. implementation strategy for the Superfund program. For Superfund cleanups at sites with many potentially responsible parties, EPA needs to take legal action against only one of the parties to recover the cost of cleanup. This party can then petition the court to bring the other potentially responsible parties into the case. This is an attempt to maximize the amount of funds that others contribute to the cost of cleanup and thereby minimize their own costs. EPA can initiate action more quickly and with less expense than if legal proceedings required thoroughly researched cases for all of the potentially responsible parties. However, this strategy produces a great deal of litigation among the potentially responsible parties to determine everyone's share of the cleanup cost. Although this strategy holds down the direct cost to the EPA, it increases the cost to private firms and the overall cost to society.

In addition to litigation among the potentially responsible parties, joint and several liability contributes to the considerable litigation with and among insurance companies. Many parties had multiple insurers over the years when they contributed toxic waste to a specific contaminated site, which leads to extensive secondary litigation between the potentially responsible parties and their insurers. This litigation can substantially increase the transaction costs and the time required to complete environmental cleanup of a contaminated site.

In the United States, CERCLA dramatically changed the probability basis of environmental risks. Judicial interpretations of the coverage of insurance contracts for environmental liability are determined by the states. They have varied greatly, causing further confusion and increasing the perception of risk. Environmental liability has greatly increased the costs and limited the availability of insurance for both corporations and municipalities (Sommerfield, 1990; U.S. General Accounting Office, 1989).

DEEP POCKETS

The litigation process under joint and several liability often shifts the focus from concern with forcing the polluter to pay for environmental cleanup to the ability to pay. Not all of the potentially responsible parties have equal assets. Often one or more have a greater capacity to pay, which is called having "deep pockets." The polluter pays principle suggests that governments should direct action toward those parties that contributed the most to the contamination. The litigation process under joint and several liability, on the other hand, often leads to financial responsibility for those who can contribute most to the cost of cleanup.

Potentially responsible parties with deep pockets argue that the Superfund program singles them out for exorbitant charges, even if they were only marginally involved in the cause of the contamination. The litigation process typically leads to the identification of insurance companies, municipalities, and large multinational corporations as having deep pockets. The insurance companies and large corporations possess considerable assets, whereas municipalities and other units of government can raise funds through their taxing powers.

PLANNING FOR ENVIRONMENTAL CLEANUP

Even with strong legal powers, the polluter pays principle will not be able to raise sufficient funds for environmental cleanup at many contaminated sites. In the United States, some states have "superlien" laws that give the state priority over all other liens to ensure that any available funds be applied to cleanup costs.

The capability of court systems to implement the polluter pays principle has serious limitations. As a practical matter, countries that base their environmental cleanup policy on this principle must supplement these funds with funds from other sources. Typically, countries supplement private party funds with special taxes on targeted industries and/or with funds from general revenues. The taxes are levied on industries that are historically associated with substances and activities that caused contaminated sites (e.g., see The Netherlands case study in Chapter 10 and the Superfund program in Chapter 8), and so in this sense adhere to the polluter pays principle.

Countries without environmental cleanup policies may be forced to deal with the issue. Such environmental liability first arose in Europe through the U.S. subsidiaries of large European corporations. There are numerous examples of European corporations buying U.S. firms and thus becoming liable for the cleanup cost of contaminated sites that they now owned in the United States (House, 1993). In the years immediately following enactment of the Superfund program with CERCLA in 1980, the implications of environmental liability and the huge cost of environmental cleanup were largely unknown. Even after the implications became known in the United States, many European firms were still operating in countries without environmental cleanup policies (see Chapter 9). Companies based in countries with well-developed policies bring these issues along with their activities and plants to other countries (see Chapter 7). Some firms may choose to invest only in countries without environmental cleanup policies to maximize short-term profits. Multinational firms may decide to only invest in countries with clear cleanup policies in hopes of a consistent and stable business environment.

MUNICIPAL LIABILITY PROBLEMS

In some countries, such as the United States, local governments have large environmental cleanup liabilities. Municipalities are liable for cleanup costs primarily for two reasons: (1) they previously sent municipal solid waste to landfills and (2) they took ownership of contaminated property because the owner failed to pay property taxes.

As discussed earlier, joint and several liability makes any party that contributed to a contaminated site potentially liable for the cost of the entire cleanup. Because municipalities have deep pockets, they face a serious risk of being held totally responsible if government cannot find other responsible parties. This scenario is not uncommon because municipalities have often contributed waste to landfills that are now contaminated and because some portion of municipal solid waste inevitably contains toxic substances. Furthermore, municipalities owned, and continue to own, landfills that are now contaminated. On the list of the worst contaminated sites in the United States, the National Priorities List, there are 230 municipal landfills in which industrial waste was co-disposed with municipal waste (Steinzor and Kolker, 1993).

Even though less than 1% of municipal solid waste is hazardous material, the Superfund program threatens municipalities with paying a much larger share of the cost of cleanup. Other responsible parties usually claim that the proportion of cleanup costs should be based on the *proportion of total waste* contributed to the landfill because no one can determine the *amount of toxic waste* that each contributed. Municipal solid waste was often the largest volume contributor to landfills that are now contaminated.

The Superfund program forces municipalities to make difficult choices. Typically, they can settle the case or risk large legal fees to defend themselves from the liability. In California, local governments have been sued for doing nothing more than issuing business licenses to private waste haulers who contract directly with citizens for solid waste collection and disposal (Steinzor and Kolker, 1993). In response, the EPA has attempted to protect municipalities from large cleanup costs, and some proposals to reform the Superfund program contain limits on municipal liability.

The second way that U.S. municipalities assume large liabilities for contaminated sites is through taking ownership of tax-delinquent properties that are later discovered to be contaminated. In some states, municipalities are required to assume ownership of tax-delinquent property, and in most states it is standard operating procedure. The goal in these cases is that the municipality may then sell the property to a developer to gain revenue, thus returning the property both to productive use and the tax rolls. Though federal regulations protect municipalities from liability for contamination on "involuntary takings" property, some states do not provide municipalities this protection.

In states that do not require the confiscation of tax-delinquent property, some cities have stopped taking such contaminated properties in lieu of taxes owed. In 1991 and 1992, the city government of Milwaukee, Wisconsin, conducted 590 site screenings to investigate contamination of suspected tax-delinquent properties. It subsequently placed 160 of the properties on its "do not acquire" list, which saved the city an estimated $30 million in potential cleanup costs (Salcedo and Dettmann, 1993).

Cities in the United States that refuse for financial reasons to take ownership of tax-delinquent contaminated properties may cause unintended consequences. Such properties exist in a state of "limbo." Even in instances where a city refuses to take ownership of the property, it may decide not to fence the property to protect the public because that action may constitute "active management" of the site and thereby make them liable for the cleanup costs. This creates a profound moral dilemma. For example, unsuspecting children may end up playing on abandoned and contaminated sites that threaten their health. This can be a serious problem in older industrial cities in the United States. In response to these and related concerns, EPA is developing guidelines on municipal liability that will clarify the legal and financial ramifications of such situations.

PRIVATE FIRMS' LIABILITY PROBLEMS

Private firms that may be liable for environmental cleanup must take steps to protect themselves. Five such actions are: (1) conduct environmental audits of their property and especially any property they consider buying; (2) increase insurance coverage (insurance and environmental cleanup liability are discussed in Chapter 6); (3) set up financial reserves (this may create problems by reducing corporate dividends); (4) record a contingent liability in a footnote to the balance sheet and hope it isn't needed (see Chapter 6 for a discussion of accounting rules); and (5) establish purchase agreements that include indemnity arrangements for purchase of the property that they may later discover is contaminated.

Environmental audits to detect contamination are becoming an increasingly common and required part of commercial or industrial real estate transactions. Audits are the most important action that private firms must take to protect themselves from cleanup liability. Several U.S. states have passed legislation that requires sellers to disclose the presence of hazardous materials. New Jersey passed the first legislation in 1983 that required disclosure of contamination and an environmental cleanup plan before an owner could sell land. Even when laws do not require disclosure or the owner of the land does not know that contamination exists, it is still wise to conduct an environmental audit. Lending institutions now require environmental audits before lending money for virtually all commercial and industrial land transactions.

Environmental audits are procedures for measuring, evaluating, or appraising the status or probability of contamination from previous uses of the property. This invariably consists of a Phase I and perhaps a limited Phase II inspection. A Phase I inspection includes a thorough walk-through of the site, as well as a document search for previous ownership and activities, to determine the possible presence of hazardous substances. A Phase II assessment includes the testing of actual material samples (soil, water, etc.) for toxic substances.

If an audit discovers contamination, the seller may take the cost of remediation into account and reduce the asking price accordingly. The environmental audit is an important part of the "due diligence" defense against liability. This defense argues that the purchaser is innocent of causing or even knowing about contamination and should be free of liability for environmental cleanup because the purchaser made all reasonable efforts to discover any contamination.

Many private firms establish financial reserves against the threat of environmental liability. In the United Kingdom in 1992, a firm set aside £734 million to pay for any environmental liability that may arise from the purchase of a subsidiary purchased for a total of only £350 million (House, 1993). The first amount was equivalent to three-fourths of the purchasing firm's 1992 profits.

Special purchase agreements vary, but often include long escrow periods and some form of indemnity. A long escrow period gives the purchaser time to search more thoroughly for contamination while the purchaser holds back a large sum of money from the seller in an escrow account.

An indemnity agreement attempts to remunerate another for loss or to protect that party against liability. Purchasers often require indemnity when purchasing a firm or property to help protect against liability for environmental cleanup. In other words, sellers indemnify purchasers against the cost of cleanup if the government should require a cleanup at some future date. In Europe, private firms selling property try to keep the indemnity below 100% of value, but in the United States, it is quite common to give unlimited indemnity.

The changing standards for liability caused by the need to raise large sums of money to pay for cleanups can be disruptive to banks. In the United States, this problem caught financial institutions off guard. They are now sensitive to dealing with contaminated sites in commercial real estate. The resolution of *United States v. Maryland Bank & Trust* (1986), and other cases in which the court found the lending institution liable for the cleanup of contaminated property obtained through foreclosure, was a wakeup call to financial institutions.

In 1990, for example, financial institutions stopped loaning money for real estate purchases on all properties in a 28-square-mile area in Tucson, Arizona. These properties were located above a large plume of contaminated groundwater and owners of the property were all "potentially responsible parties" that might be held liable for the cleanup of the Tucson Airport Superfund site. After several months, the secondary markets adjusted to the risk of liability and resumed making loans.

Financial institutions now routinely use a due diligence process to scrutinize potential commercial loans for contamination during the underwriting process. The U.S. EPA has issued guidance that they will not hold banks liable for properties they did not own or actively manage. Environmental liability in Europe continues to spread confusion and uncertainty among banks, where liability has changed standards for capital adequacy, threatened insurers with insolvency, and caused investors to change their plans about acquisitions.

CONCLUSIONS

All countries that have recognized the toxic contamination and environmental cleanup problem use laws to attempt to prevent new contaminated sites and force the cleanup of existing sites. In theory, economic systems could prevent the problem without government regulation, but, in practice, laws and regulations are needed.

Using laws to address this problem is a command and control form of environmental policy. Such policy can be effective, but may not be the economically most efficient means to achieve the goal. Increasingly, countries are experimenting with manipulating economic incentives to achieve environmental protection without the need for command and control environmental policy. For example, it may be possible to establish sufficient economic incentives to induce developers to buy contaminated sites and perform the environmental cleanup. Legal approaches are becoming more common for most countries, until analysts can design and prove that alternatives are effective.

It is still unclear whether the use of powerful legal tools can effectively force the private sector to fund environmental cleanup. At one extreme is the case of the United States, where the government is willing to use powerful liability techniques to pay for the high cost of cleanup. Many citizens think that some of these techniques are unfair. At the other extreme is Germany's experience in completing the environmental cleanup of toxic contamination sites in the former East Germany (see Chapter 9); they decided to use general government revenues to pay for the cleanup. Most countries are likely to use legal powers that lie somewhere between these two examples.

References

House, R. 1993. Balance-sheet poison: Corporations and their creditors find there is no cover to run for as environmental liability comes to Europe. *Institutional Investor,* Int. Ed. **18**(8), 23–28.

Oswald, L. J. 1993. Strict liability of individuals under CERCLA: A normative analysis. *Boston Coll. Environ. Aff. Law Rev.* **20,** 579–637.

Russel, M. E., W. Colglazier, and M. English. 1991. *Hazardous Waste Remediation: The Task Ahead.* Knoxville, TN: Waste Management Research and Education Institute.

Salcedo, R. and D. Dettmann. 1993. Cities wrestle with abandoned properties. *Pollut. Eng.* June 1, 76–78.

Sommerfield, F. 1990. Going bare. *Institutional Investor,* **24**(March), 99–102.

Steinzor, R. and D. Kolker. 1993. Superfund liability dumped onto local governments. *Gov. Finance Rev.* **9**(4), 11–13.

U.S. General Accounting Office. 1989. *Hazardous Waste: The Cost and Availability of Pollution Insurance.* Washington, DC: U.S. Government Printing Office.

Wilkerson, W. and T. Church. 1989. The gorilla in the closet: Joint and several liability and the cleanup of toxic waste sites. *Law Policy* **11**(4), 425–449.

CHAPTER 6

ECONOMIC, SPATIAL, AND SOCIAL ASPECTS OF CONTAMINATION AND ENVIRONMENTAL CLEANUP

INTRODUCTION

The economic issues in contamination and environmental cleanup stem from the large numbers of contaminated sites, the threat to ecosystem health, the human health risks, and the expense of cleanup. In addition, contaminated sites are distributed in a distinctive spatial pattern and cause serious social problems. An understanding of the nature of these problems is essential when considering policies to prevent and clean up additional contaminated sites and when developing regulatory institutions to implement these policies.

Toxic and radioactive substances have the potential to move through environmental pathways from contaminated sites; therefore, they pose a risk to the health of human populations and ecosystems in a potentially wide region. This potential mobility can result in economic costs to the whole society and not just the immediate geographic area where the contaminated site is located.

This chapter discusses the economic frameworks in which the contamination and environmental cleanup problem exists. It examines how contamination and cleanup affect banking systems, real estate markets, and insurance markets. The spatial and social effects caused by the concentration of contaminated sites will also be considered.

ECONOMIC PROBLEMS OF CONTAMINATED SITES

Both market and state-controlled economic systems have produced large numbers of contaminated sites that need environmental cleanup. Free markets have failed to provide sufficient incentives to industry, agriculture, mining, and other activities to keep the use and disposal of toxic substances from escaping into the environment. State-controlled economic systems have also failed to prevent environmental contamination, even though they had the means to avoid the problem (see Chapter 12).

The external diseconomies arising from contaminated sites in market economies are a result of market failure. Because private prices do not reflect external costs or benefits, markets fail to provide sufficient incentives to carefully control the use and disposal of toxic substances. Although the market assigns no cost to the disposal of hazardous wastes on land, such disposal produces large negative "spillover effects" on society. By ignoring the nonmarket benefits of land protection and the costs of spillover effects, firms overutilized land disposal and created large numbers of contaminated sites. This resulted in the abandonment of many sites because cleanup is so expensive.

Because greenfield sites—those free of contamination—are more attractive than contaminated brownfield sites, market forces lead new development to greenfield sites. Suburban municipalities offer developers many subsidies to attract them to greenfield sites, whereas developers considering brownfield sites face the risk of liability for environmental cleanup costs. This is referred to as a "lumpy investment" situation, which deters many purchasers. Market failure often occurs when there are lumpy investment requirements.

Because markets do not accurately reflect the true social costs for the release of toxic substances to the environment, government intervention to remedy the cleanup problem is necessary. When industries lack sufficient market incentives, they will likely fail to prevent toxic substances from escaping and becoming a threat to humans and ecosystems. Unless there is strong and enforceable policy intervention, markets will continue to ignore the benefits of protecting the public and environment.

In addition to government policies to protect public health and the environment, countries need restrictions on the international trade of hazardous wastes. Even stalwart advocates of free markets and free trade recognize that hazardous wastes require government market intervention (World Bank, 1993, p. 67).

Yet some people question whether the existence of contaminated sites is a problem. They argue that if the contaminated property has sufficient value, real estate markets will find someone to buy the property, remediate the contamination, and redevelop the site. They also argue that if markets do not encourage the remediation and redevelopment of sites, then the land should remain contaminated and abandoned.

From this economic perspective, government intervention to force remediation of contaminated sites is an inherently inefficient use of funds that the free market would otherwise allocate to economically more productive activities. Accordingly, such government policy is likely to be inefficient and ill-advised. The only exceptions to this generalization are those few instances where a proven public health problem exists (see Chapter 3).

Others believe that these economic arguments are mistaken and shortsighted. First, there are several reasons why contaminated sites are a problem. Toxic contaminants may now or in the future migrate by poorly understood environmental pathways and expose people to single chemicals or mixtures of chemicals with unknown health implications. Second, contaminated sites seem to attract an array of socially undesirable activities that become a social cancer on the urban environment. This produces a secondary negative externality that further hinders the real estate market from operating efficiently. There are also spillover effects on neighboring properties, whose values may be reduced.

Finally, there are intergenerational inequities associated with leaving land contaminated and abandoned. If sites are remediated and redeveloped, they could produce a long stream of benefits from the productive use of the site. If the site remains contaminated, entrepreneurs will not redevelop the site, because greenfield sites will almost always be more attractive when suburban subsidies exist. Land is always in limited supply, so lack of redevelopment means that future economic growth may be diminished. Intergenerational effects also arise because toxic chemicals on land often destroy ecosystems. Degraded ecosystems may differ in value, but all ecosystems have inherent value for present and future generations. This is the same logic as the moral and economic arguments for saving rain forests or endangered species.

Although the standard free market argument asserts that markets should solve this problem, economic reality is more complex. The factors that prevent private markets from solving the contamination and environmental cleanup problem include: subsidies to greenfield development, lumpy investments to clean up and redevelop contaminated sites, negative spillover effects, and uncertainties about future health effects.

ECONOMIC DISTORTIONS

The ecosystem and human health risks of toxic substance contamination and the great cost of environmental cleanup create economic distortions.

Because of liability, individual firms, especially those that use any toxic substances, must set aside large sums of money. For example, in the United States in 1994, ARCO announced that environmental cleanups for which it was responsible could cost $1 billion more than the $648 million it already had set aside for that purpose (*Environmental Damage Valuation,* 1994). Company stock values may be affected because governments are now forcing private firms to publicly release information regarding their environmental liabilities. In the United States, accounting rules require that companies include detailed data on known environmental liabilities. In Europe, comparable accounting rules do not exist and liability information is generally not available. This means that European firms may be currently overvalued. This asymmetrical treatment may in turn bias investment patterns in an increasingly global marketplace.

The impacts of environmental cleanup on the banking, real estate, and insurance industries illustrate some of the distortions that influence economic systems. Environmental policies attempt to minimize these economic distortions, as they simultaneously address the environmental and human health problems.

Effects on Banks

The United States is the clearest example of a country where legal liability for environmental cleanup affects banks. In an attempt to force the private sector to pay for cleanups, the United States extends legal liability to extreme lengths, to the point where even nonpolluters can be forced to pay for the cleanup. This occurs when the actual polluter is not found or cannot pay (see Chapters 5 and 8). In some instances, banks are held liable, often when a property owner defaults on a loan and a bank takes ownership of the property held as collateral.

The legal complications and financial risks resulting from even tangential involvement with contaminated sites have reduced the available financing for these sites. Lenders may be unwilling to provide loans for several reasons: (1) concern about their own liability, (2) the reduced collateral value of the land if it is found to be contaminated, and (3) the ability of the property owners to repay the loan if they must pay for remediation of the contaminated property. Local banks have become increasingly dependent on the legal advice of environmental attorneys who specialize in minimizing a bank's liabilities. These decisions often prevent banks from applying their economic influence and knowledge of real estate markets and therefore act to limit redevelopment of remediated sites. To protect the banking and real estate industries from such uncertainties, the U.S. federal and state governments are attempting to define and limit lender liability for environmental cleanup.

Effects on the Real Estate Industry

Environmental liability affects the real estate industry because ownership of contaminated property can result in liability for the cleanup. This has a great impact on the real estate industry in countries that rely on the polluter pays principle, whereby the private sector is forced to pay for environmental cleanups.

The 1980 Comprehensive Environmental Response, Compensation, and Liability Act (CERCLA), had the general effect of lowering property values near a contaminated site. In a New Jersey study, the sale prices of homes in 77 communities with hazardous waste Superfund sites increased significantly less than those in control communities (Greenberg and Hughes, 1992). Several studies found that the price for a home increased somewhere between $3310 and $14,200 for each mile away from a hazardous waste site (Kohlhase, 1991; Smolen *et al.*, 1992). Another study in New Jersey found that if the number of hazardous waste sites in a municipality decreases by one, the median property value increases by an average of 2% (Ketkar, 1992). Some experts estimate that declines in real estate values from toxic substance contamination total $2 trillion (Mundy, 1992). Other researchers estimate that the "stigma" loss to the value of property suspected to be contaminated, or even for property near a contaminated site, can be 30–35% of the land value (Patchin, 1994). This stigma loss occurs regardless of actual contamination (McCelland *et al.*, 1990).

In countries that use the polluter pays principle, the real estate industry has responded by strict adherence to "due diligence" (see Chapter 5), in which a party in a real estate transaction thoroughly investigates the property to determine the presence of any contamination. The party can then defend against future liability by claiming that they took all reasonable steps regarding the condition of the property. If contamination is later discovered on the property, the government will treat the party who exercised due diligence as an innocent owner who is not liable for cleanup costs. This is critically important, because even conscientious attempts to investigate for contamination may fail. It is estimated that Phase One environmental audits miss 40% of all contamination (Sibley, 1992).

An environmental assessment can be part of a legal defense against liability based on due diligence. This defense must prove that the firm: (1) is an innocent owner, (2) was not involved in the original contamination, (3) diligently avoided creating new contamination, and (4) has a record that contamination was not identified at the time the site was purchased, despite an appropriate environmental assessment. Even though the due diligence defense is clearly the best strategy for any party involved in redevelopment efforts, it may not be sufficient to avoid liability for a contamination cleanup under CERCLA.

Brownfield Sites

Brownfield sites are sites that no one is presently using, but that previously were used. They can be a particular problem if they were formerly industrial sites, are located in run-down urban areas, and are contaminated. Environmental laws deter parties interested in redeveloping contaminated brownfield sites for two reasons: (1) CERCLA and the Superfund Amendment and Reauthorization Act (SARA) impose potential liability that may lower property values more than the actual cost of environmental cleanup (Page and Rabinowitz, 1993) and (2) the threat of delays and higher costs resulting from unpredictable cleanup policies produce a stigma effect (Arrandale, 1992).

Individual states within the United States also have programs to overcome the legal difficulties and to promote reuse of contaminated sites and brownfield sites. State efforts protect innocent owners from liability by providing covenants not to sue, clarifying the lender's liability, and attempting to streamline the state's regulatory processes (U.S. General Accounting Office, 1995). Several states have passed legislation that requires sellers to disclose the presence of hazardous materials. However, state programs do not protect those involved with contaminated sites from liability under federal law.

Despite efforts by the U.S. government, the real estate industry has great difficulty redeveloping contaminated sites, even after an environmental cleanup. The Netherlands and other European countries have been more successful in redeveloping contaminated brownfield sites (see Chapter 9). In the United States, developers will complete cleanup of a contaminated site and return it to productive use only under special circumstances (Page and Rabinowitz, 1994).

Effects on the Insurance Industry

The environmental cleanup problem has also affected the insurance industry in two ways: (1) by producing large claims on existing policies and (2) by creating proposals for new insurance trust funds to pay for environmental cleanup. The insurance industry has been most directly affected because in many countries the courts have ruled that existing insurance policies must cover the costs of cleanup.

Historically, insurance companies had written policies providing insurance against sudden events such as accidental spills or releases. However, they had not considered the costs of environmental cleanup of gradual pollution when writing these policies. Despite the protest of the insurance industry, the courts decided that these policies did in fact cover gradual releases of toxic substances. This meant that policy holders who did not detect pollution until as much as 30 years later could still make claims, because they had insurance when the events took place.

In response to these court decisions, insurance practices now vary in different countries. In the United States, CERCLA eliminated the tradition of long-term insurance relationships, often of 30–40 years. In the United Kingdom, with only a few exceptions, insurers have collectively refused to underwrite anything but sudden and accidental liability policies since 1991. France, Italy, and The Netherlands have insurance pools, but coverage for contaminated sites is not complete.

Many European countries have different legal systems and therefore have experienced different impacts on their insurance industries. For example, most European countries do not use juries in civil cases. As a result, court awards are less "erratic" than in the United States.

German insurance companies still offer insurance for "gradual" as well as sudden and accidental coverage. This is probably because the German legal system discourages litigation and high transaction costs, which are common in the United States and United Kingdom. The German style of litigation uses special judges, expert witnesses, and a loser-pays-the-winner's costs provision that provide a strong incentive to settle a case rather than litigate. An example of this difference is the experience of the German company Haftpflichtverband der Deutschen Industrie VVaG in Hanover. This firm experiences 25,000 pollution claims annually in Germany, and 150 annually in the United States. Yet only 2% of this company's claims end up in German courts, whereas it litigates 93% (140 of 150) of the claims in the United States (Ladbury, 1993).

In some countries, the idea of using insurance trust funds to pay for all environmental cleanup costs is under consideration. These countries view this proposal as an alternative to using liability systems to force the private sector to pay for cleanup. For example, in 1994, the U.S. government proposed to raise $3.1 billion over a five-year period through special taxes on property and casualty insurers to finance an "environmental insurance resolution fund" (Roberts, 1994). It was hoped that such a broad-based tax could raise sufficient funds to create an effective insurance fund. Such a system would reduce transaction costs and provide policy holders a portion of their cleanup costs in return for an agreement not to sue their liability insurers.

There are other innovative proposals for using insurance companies to help finance environmental cleanup. In New York City, a Gowanus community group is working with a local insurance company to form a company called Remedial Capital Corporation. This corporation would buy contaminated sites, supervise the cleanup, and underwrite insurance that would indemnify all future developers (Roberts, 1994).

ENVIRONMENTAL REGULATION

The risks to human and ecosystem health posed by contamination represent costs to a country. These include future human health costs, environmental damages, and the opportunity cost of not having all of these contaminated sites in productive use. Most countries have decided that a policy to clean up the most serious contaminated sites will cost less over the long term than taking no action.

The effects of environmental regulations on a country's economic performance can vary. Some environmental policies will be inefficient because the sum of the costs of implementing the policy will exceed the sum of the benefits of the policy. Other policies may be inefficient because they do not do enough to minimize a specific environmental problem. In this case, additional environmental protection would have produced greater benefits than the cost of the additional protection. One of the difficulties in policy making is that one can never accurately predict or measure all the future costs and benefits. Once a government implements a policy, it can gather information that provides a better sense of the benefits and the costs. However, even after a policy has been in place for many years, these may still be difficult to measure.

In the United States, researchers have studied the economic effects of the dramatic environmental regulations first enacted in the early 1970s. By the mid-1980s, the overall effect of environmental policies on employment was mildly positive, meaning that the policies caused a modest increase in employment. The U.S. gross national product (GNP) increased in the initial period 1970–1978 as the government implemented regulations. However, by 1986, researchers estimated that GNP was more than $24 billion less (a change of less than 1%) than it would have been without environmental regulations (Seneca, 1987, p. 360). Others have estimated that a small but detectable part of the decline in the U.S. productivity growth is due to environmental regulations (Haveman and Christainsen, 1981). Slower productivity growth decreases the international competitiveness of U.S. markets and is directly related to the deterioration of the trade balance (Seneca, 1987, p. 361; Congressional Budget Office, 1985). There is also evidence that some long-term deterioration in the U.S. trade balance is attributable to environmental controls (Seneca, 1987, p. 361).

Proponents of environmental regulation claim that these studies do not adequately measure the benefits of environmental protection. They claim that economists measure GNP (or gross domestic product GDP), in ways that underestimate the benefits of environmental regulations, because GNP or GDP measures quality of life and ecosystem health benefits only indirectly, if at all, and probably with a substantial lag period.

The prevailing view is that a short-term decline in the rate of economic growth is a worthwhile investment to reap the long-term benefits of a clean environment. Because wealth-producing economic activities are increasingly free to locate anywhere on the planet, a clean environment will be increasingly important in attracting these activities. Firms are likely to place more value on a clean environment and the heightened quality of life that accompanies it in the future. This is especially true because the cost of future environmental cleanup will rise as toxic contaminants migrate from their original location to contaminate greater amounts of land and groundwater.

SPATIAL IMPLICATIONS

Most of the industrial contaminated sites that require environmental cleanup are located in urban areas, although land contaminated by mining and the military is usually located in rural areas. Until recently, the concentration of industrial activity in urban areas was one of the main characteristics of industrial societies. This spatial concentration allowed the access to labor and the physical infrastructure that industry required. Today, economic activities do not need to be as geographically concentrated as in the past. The changing spatial patterns of economic activity are an important component of environmental cleanup policies.

The overall trend for all countries is toward greater urban concentration. The world is likely to be more than 50% urban by the year 2000. The United States was 33% urban in 1950 and 77% urban in 1990 (Flynn, 1992; Nathan, 1992). Yet the developed countries are becoming both more urban and more decentralized at the same time, as the dominant urban pattern is changing. In the most highly developed countries, the trend is for decentralization from the old urban centers to the periphery of metropolitan areas at ever-increasing distances.

This pattern of a dispersed form of increased urbanization is most pronounced in the United States, but is a trend in all the developed countries. Advances in transportation and information processing technologies are driving these changes in spatial form. The U.S. industrial restructuring is characterized by the movement of jobs and population away from the center of urban areas to the periphery. Some consider this emerging spatial pattern of American urban areas to be a continuation of the process of suburbanization. Others see the process as creating a new urban form called "edge city" (Garreau, 1991) or "citistates" (Pierce *et al.*, 1993).

The globalization of the world's economy interacts with the issues of contaminated sites, environmental cleanup, and redevelopment of land in important ways. Free market economic forces, which are now driving the process of globalization, caused the contamination that is a problem today and produced large numbers of abandoned industrial properties in urban areas.

In the United States and other mature economies, often referred to as postindustrial countries, economic activity is changing dramatically. Some analysts refer to this shift as the "informational" mode of development, in which the pivotal industries such as semiconductors, computers, genetic engineering, and telecommunication are the driving force (Castells, 1989). Others argue that changes in the financial industry from a services concentration to a commodities concentration, combined with the growth in international investing, are causing the globalization of the world economy (Sassen, 1991). These economic forces have essentially decentralized economic growth to sites throughout the world and to the periphery of metropolitan areas. This decentralization produces increased and more visible differentiation among income groups. These changes also leave contaminated brownfield sites in urban areas with limited prospects for private sector redevelopment. Such sites are not only unproductive, but are not producing tax revenues that cities could use in an attempt to solve serious environmental, economic, and social problems.

The prospect of redeveloping or reusing presently contaminated brownfield sites has spatial implications. Where countries do not remediate and reuse contaminated sites, holes in the urban fabric appear and social problems follow. Many people believe that this is an issue of "environmental justice" because contaminated sites needing cleanup are concentrated where the poor and minorities live. Political and economic forces tend to locate locally unwanted land uses (LULUs), such as industries using toxic substances, waste-processing facilities, and landfills, in locations where the local citizens are neither politically nor economically powerful. The poor are no match for the wealthiest and most politically powerful segments of society who directly benefit from the current location of LULUs and do not want these facilities near their homes. Governments have an obligation to represent the interests of the urban poor and ethnic minorities to counter such environmental injustice aspects of the cleanup problem (World Bank, 1993, p. 83).

Most of the effects of global economic restructuring and their interactions with environmental cleanup occur in all the industrialized countries. A combination of national policies and the present stage of economic transformation makes the spatial implications of these problems most severe in the United States. Though the United States is a world leader in environmental cleanup, it trails far behind the Western European countries in returning these remediated brownfield sites to productive use.

In Europe, reuse of remediated brownfield sites is likely to happen sooner and more often than in the United States because of differences in land markets. European countries have greater control of urban boundaries and a greater degree of protection from development of open space (see Chapter 9). Together, these more potent land use controls increase the obstacles to developing "greenfield" sites on the periphery of metropolitan regions and encourage the development of "infill" sites. Many redevelopment sites within urban areas are likely to be brownfield sites and some of them are likely to be contaminated with toxic or radioactive substances before the government forces an environmental cleanup.

SOCIAL PROBLEMS

The changing spatial organization of the urban landscape has important implications for society. Research in France, the United Kingdom, and the United States found that global economic restructuring was causing the marginalization of ethnic minorities in all three countries (Keating, 1993). In the United States, the accelerating decentralization of economic activity is producing increasingly significant inequalities (Goldsmith and Blakely, 1992). Since the early 1970s, industrial restructuring has caused greater disparities in wealth. Families with lower income became worse off while the income of wealthy Americans increased (Rivlin, 1992). Upper- and middle-income whites moving out of the metropolitan core cause growth in the peripheral portions of urban areas. The old central cities of the United States, where most industrial contaminated sites are located, are now primarily the home of lower-income groups, largely nonwhite and often recent immigrants.

Countries that fail to redevelop contaminated brownfield sites deprive urban areas of desperately needed jobs and the positive economic multiplier effects of redevelopment, and such neglect may contribute to public health problems. Besides contamination resulting from previous activities, abandoned contaminated sites may attract illegal activities such as additional dumping of hazardous waste, break-ins, fires, drug dealing, and other unwanted activities.

Researchers refer to brownfield and other derelict sites in urban areas as TOADS (Temporarily Obsolete Abandoned Derelict Sites), and they often become epicenters of socioeconomic malaise and public health problems (Greenberg *et al.*, 1990, 1992). Blighted neighborhoods encounter an array of problems that residents perceive more intensely and differently than do outsiders looking at these problems (Greenberg and Schneider, 1996). Once developers identify a neighborhood as blighted, they will rarely invest there (Goetze, 1976; Clay and Hollister, 1983).

In the developed countries, there are calls for new policy to counteract the spatial trends and their negative social impacts that result from global economic restructuring. Some of these proposals clearly address the spatial implications (Grigsby and Godshalk, 1993), but there is currently little political support in the United States for these ideas. Some proposals suggest eliminating numerous federal programs and shifting the responsibility to the states, where bottom-up community efforts and sensitivity to local conditions can produce effective economic development (Rivlin, 1992). Other research suggests that it is difficult to distinguish whether economic restructuring or local unemployment is the cause for specific socioeconomic problems (Adams *et al.*, 1993).

In contrast to an explicit urban agenda, a common response is to call for increased economic activity and productivity. Many argue that the national economy must be growing to effectively address social problems. However, the forces driving economic globalization and urban changes are likely to increase the inequality unless countries enact strong policies to alter the existing trends.

The changing urban landscape in the developed countries contributes to environmental problems in addition to those associated with cleanups of contaminated sites. The central cores of metropolitan areas that house the poor and minorities often have the greatest burdens of pollution and the fewest resources to lessen exposure to that pollution. This is one form of the problem known as "environmental racism" whereby society exposes those least able to defend themselves to the greatest concentrations of contamination.

Contamination from old industrial sites is one source of exposure, but air and water pollution from many other sources is also present. Furthermore, the financial resources to adequately maintain the aging physical infrastructure are too often insufficient. At the same time, the trend toward decentralization of economic activity creates urban sprawl that consumes open space and habitat on the periphery of cities, causing further pollution.

CONCLUSIONS

The toxic contamination and environmental cleanup problem is a dauntingly serious economic problem for there are nearly countless contaminated sites and cleanup is very expensive. This problem is relatively new, and the economies of different countries are adjusting in a variety of ways and at different speeds. The extent of the economic and social problems caused by contaminated sites is so great that the economically most developed countries are already developing policies to ameliorate the problem.

The greatest economic issue facing countries is deciding how much they should spend on environmental cleanup. Countries have many competing demands for limited financial resources, and because cleanup is so expensive and there are so many contaminated sites, this problem threatens to seriously deplete available revenues. Another issue is how much can countries spend on environmental cleanup without damaging their economic competitiveness? Each country must examine its social and economic priorities as it attempts to answer these difficult questions.

References

Adams, C., L. Russell, and C. Taylor-Russell. 1993. Land use policy responses to economical restructuring. *Land Use Policy* **10**(2), 151–172.

Arrandale, T. 1992. Developing the decontaminated city. *Governing* **3**(3), 44–47.

Castells, M. 1989. *The Informational City.* Oxford: Basil Blackwell.

Clay, P., and R. Hollister (eds.). 1983. *Neighborhood Policy and Planning.* Lexington, MA: Lexington Books and Heath.

Congressional Budget Office. 1985. *Environmental Regulation and Economic Efficiency.* Washington, DC: Congress of the United States.

Environmental Damage Valuation and Cost Benefit News. 1994. ARCO increases cleanup estimates by $1 billion. Vol. **1**(1), Summer.

Flynn, R. L. 1992. Rebuilding America's cities together. Draft, cited in Stegman, M. A. 1993. National urban policy revisited. *N. C. Law Rev.* **71,** 1737–1777.

Garreau, J. 1991. *Edge City: Life on the Frontier.* New York: Doubleday.

Goetze, R. 1976. *Building Neighborhood Confidence.* Cambridge, MA: Ballinger.

Goldsmith, W. W., and E. J. Blakely. 1992. *Separate Societies: Poverty and Inequalities in U.S. Cities.* Philadelphia: Temple University Press.

Greenberg, M., and J. Hughes. 1992. The impact of hazardous waste Superfund sites on the value of houses sold in New Jersey. *Ann. Reg. Sci.* **26,** 147–153.

Greenberg, M., and D. Schneider. 1996. *Environmentally Decimated Neighborhoods: Perceptions, Policies, and Realities.* New Brunswick, NJ: Rutgers University Press.

Greenberg, M., F. Popper, and B. West. 1990. The TOADS: A new American urban epidemic. *Urban Stud.* **25**(3), 435–454.

Greenberg, M., F. Popper, B. West and D. Schneider. 1992. TOADS go to New Jersey: Implications for land use and public health in mid-sized and large U.S. cities. *Urban Stud.* 29(1), 117–126.

Grigsby, J. E., and D. R. Godshalk. 1993. Report: Reshaping America's urban agenda. *J. Plann. Educ. Res.* **13**(1), 60–66.

Haveman, R. H., and G. B. Christainsen. 1981. Environmental regulations and productivity growth. In H. M. Peskin, P. R. Portney, and A. V. Kneese (eds.), *Environmental Regulation and the U.S. Economy.* Baltimore, MD: Johns Hopkins University Press.

Keating, M. 1993. The politics of urban development: Political change and local development politics in the United States, Britain, and France. *Urban Aff. Q.* **28**(3), 373–396.

Ketkar, K. 1992. Hazardous waste sites and property values in the state of New Jersey. *Appl. Econ.* **24,** 647–659.

Kohlhase, J. 1991. The impact of toxic waste sites on housing values. *J. Urban Econ.* **30,** 1–26.

Ladbury, A. 1993. Multinational risk management: European insurer touts approach to cleanups. *Bus. Insur.* October 25.

McCelland, G., W. Schultz, and B. Hurd. 1990. The effect of risk belief on property values: A case study of a hazardous waste site. *Risk Anal.* **10,** 485–497.

Mundy, W. 1992. *Appraisal J.* April.

Nathan, R. P. 1992. *A New Agenda for Cities.* Ohio: Ohio Municipal League Educational and Research Fund.

Page, G. W., and H. Rabinowitz. 1993. Groundwater contamination: Its effects on property values and cities. *J. Am. Plann. Assoc.* **59**(4), 471–479.

Page, G. W., and H. Rabinowitz. 1994. The potential for redevelopment of contaminated greenfield sites. *Econ. Dev. Q.* **8**(4), 353–363.

Patchin, P. 1994. Contaminated properties and the sales comparison approach. *Appraisal J.* **63**(3), 402–409.

Pierce, N. R., C. W. Johnson, and J. S. Hall. 1993. *Citistates.* Washington, DC: Seven Locks Press.

Rivlin, A. M. 1992. *Reviving the American Dream: The Economy, the States, and the Federal Government.* Washington, DC: Brookings Institution.

Roberts, S. 1994. Insurers square off on Superfund reform. *Bus. Insur.* June 13.

Sassen, S. 1991. *The Global City: New York, London, Tokyo.* Princeton, NJ: Princeton University Press.

Seneca, J. 1987. Economic issues in protecting public health, pp. 351–377. In M. Greenberg (ed.), *Public Health and the Environment.* New York: Guilford Press.

Sibley, G. E. 1992. Environmental insurance. *Urban Land* July, 36–37.

Smolen, G., G. Moore, and L. Conway, 1992. Economic effects of hazardous chemical and proposed radioactive waste landfills on surrounding real estate values. *J. Real Estate Res.* **7**(3), 283–295.

U.S. General Accounting Office. 1995. *Community Development: Reuse of Urban Industrial Sites* (GAO/RCED-95-172). Washington, DC: U.S. Government Printing Office.

World Bank. 1993. *World Development Report 1992: Development and the Environment.* New York: Oxford University Press.

CHAPTER 7

CONTAMINATION AND ENVIRONMENTAL CLEANUP IN DEVELOPING COUNTRIES

INTRODUCTION

In the developing countries, the forces of international economic change are fueling rapid industrial growth, especially in firms that use toxic materials, and the quantities of toxic wastes produced will grow as their economies continue to develop. Contamination is also increasing, because controls on hazardous waste disposal in the more economically developed nations have made it cheaper for firms to ship their toxic materials and wastes overseas, in both legal and illegal ways. Production and disposal cost differentials also create international flows of toxic wastes to developing countries. Contamination will be especially severe in those countries that are slow to develop and implement effective policies to prevent the creation of contaminated sites.

Efforts to develop environmental cleanup policy must compete with other pressing problems, such as health, economic, social, and other environmental concerns. It is difficult to determine the priority for action regardless of the country or society, and contamination and cleanup may not be the greatest threat in some developing countries.

Risk assessment is often used to determine the relative priorities of problems, but they commonly fail to include the risks of contaminated sites. In particular, the threat of contamination is often ignored as a distinct risk, and it is notoriously difficult to measure the future health risks of contaminated sites. For example, a risk assessment for Bangkok, Thailand, found the risk of exposure to lead and other metals to be high but the risk of hazardous waste disposal to be relatively low (Panayotou, 1991, p. 115). Because contaminated sites produce health risks that are difficult to measure and may not arise until far in the future, these assessments are imperfect tools for ranking environmental problems (see Chapter 3).

It is important to note that toxic pollution becomes a serious problem early in the economic development process for even low-income countries. Furthermore, in the 1990s, international economic trends are compressing the classic stages of economic development; the industrial structure of developing countries is changing rapidly and the production of industrial chemicals is accelerating. For these reasons, low-income countries must start developing policies to prevent contamination early in the development process. Having policies in place before the rapid stage of industrialization begins will correct market distortions before a serious contamination problem develops. In addition, predictable environmental policy is assuring to the private sector, which wants to be able to predict their costs and avoid future environmental liabilities. Effective policies can contribute to the efficient organization of industrial and economic structure and need not reduce overall economic growth.

The enactment of policies to prevent contamination and to remediate existing contaminated sites can be cost-effective and can complement broader economic development policies, such as providing incentives for the newest and least-polluting technologies. Developing countries can learn from the experiences of the OECD countries by facing their problems early and by developing cost-effective policy to address the widely identified complexities and cost of environmental contamination and cleanup.

ECONOMIC GROWTH AND CONTAMINATED SITES IN DEVELOPING COUNTRIES

Differential economic advantages distribute the production of toxic materials and the disposal of toxic wastes around the globe. However, the costs that determine these comparative advantages do not include all environmental costs. Therefore, government policies must encourage private firms to responsibly use, produce, and dispose of toxic substances and force them to include in their accounting the costs of leaks and spills of toxic materials and the cost of safe disposal. At present, few developing countries have effective policies to prevent or remediate contaminated sites.

Economic growth in the developing countries is likely to produce substantial increases in the use of toxic materials and the disposal of toxic wastes. This occurs because countries can achieve economic development through growth in manufacturing and especially growth of smokestack industries. Many of these industries are pollution intensive and use large quantities of toxic materials. In fact, the trend of increasing emissions of toxic substances as GDP grows has remained constant for the past three decades (Lucas, 1992).

Stricter regulation of toxic substances in the OECD countries may be displacing associated industries to the developing countries, although this effect may be insignificant because emission controls represent a small fraction of operating costs. Industrial chemical production, the single greatest source of toxic pollutants, grows at 7.5% per year in the developing countries and at only 2.5% in the developed countries (Panayotou, 1991). Some empirical research reports that the poorest economies have the highest growth in industries that use toxic substances (Lucas *et al.,* 1992).

Some analysts suggest a pattern in the use of toxic materials as countries progress through the stages of economic development: toxic pollution rises faster than economic output at low levels of income, and then proportionally declines in intensity at more advanced stages of development (Lucas *et al.,* 1992; Forrest, 1995). This argument suggests that less developed countries have cost advantages that promote the growth of such industries. As these countries become more developed and achieve higher income levels, their citizens will demand improved environmental quality. At later stages, governments will also have the administrative capacity to effectively implement environmental policy.

Economists call this pattern an "inverse U-shaped relationship" between GDP per capita and the total toxic pollution from manufacturing relative to GDP, or an "environmental Kuznets curve." The turning point in emissions per unit of GDP appears to occur at a relatively high per capita income level (Lucas, 1992).

It is important to note that this inverse U relationship does not imply that the gross amount of toxic pollution decreases at higher income levels, but that faster rates of growth in other sectors of the economy, for example, the service sector, influence the overall balance. That is, the volume of toxic pollution is still increasing. When the composition of just manufacturing industries is considered, it actually becomes increasingly oriented to toxic substance production as development continues. Therefore, the quantity of toxins released to the environment, where they often create contamination, increases with higher per capita income levels. Thus increasing GDP growth rates by themselves will not solve the problem.

As developing countries increase their industrial production, the toxic pollutants that create contamination will become more widely used. There will be more contaminated mine spoils, more herbicide and pesticide production and use, more oil spills and leaks from extraction, transportation, and refining activities, more toxic material leaks and spills, and more toxic waste disposal. Countries that produce coal, oil, and gas have higher levels of toxic emissions than other countries (Lucas, 1992). Producing these raw materials seems to encourage industries that manufacture industrial chemicals, which are the industries that are the greatest source of toxic substances emissions. With rapid industrial development, the contamination and cleanup problem can quickly progress from a few isolated cases to a major problem.

CONTAMINATION AND ENVIRONMENTAL CLEANUP PROBLEMS IN URBAN AREAS

Urban areas are the location of many of the most severe contamination and cleanup problems (see Chapter 6). The large labor supply, easy access to markets, superior infrastructure and services, and the proximity of other industries compel industry to locate in cities. Contamination in urban areas is a problem throughout the world, but is especially important in developing countries where both urban areas and contamination are growing rapidly.

Cities are the engines of economic growth. Urban productivity is essential to economic development, with 80% of GDP growth in developing countries expected to originate in cities and towns (Bartone *et al.*, 1994). Research shows that industries located in a small number of metropolitan areas generates 60 to 80% of industrial value added in developing countries (Panayotou, 1991, p. 7). In Thailand, 75% of the hazardous waste generating industries are located in Bangkok, where they dispose of most waste without treatment in public places (Panayotou, 1991, p. 66). Jakarta, the capital of Indonesia, has no hazardous waste treatment facilities, yet produces 1.4 million tons of hazardous wastes per year (Panayotou, 1991, p. 109).

The population density and nearby location of residential districts can expose large numbers of people to toxic chemicals. Accidental spills and leaks are inevitable, and toxic chemicals from contaminated sites that percolate into groundwater supplies expose many city dwellers to health risks. Local governments should develop emergency preparedness and response plans for hazardous material accidents to minimize human exposure (United Nations Environment Programme, 1988, 1992; Organization for Economic Cooperation and Development, 1991; World Bank, 1985).

Often the poorest segment of an urban population receives the highest exposure to toxic chemicals. They commonly live next to industries using toxic chemicals or near marginal land where industries dump toxic chemicals. In developing countries, many poor people live near landfills and open dumps, where thousands of people scavenge the refuse for recyclable materials. In Manila, the Philippines, 20,000 people live on land surrounding a garbage dump known as Smoky Mountain and many scavenge there (Bartone *et al.*, 1994). Waste collection often mixes hazardous wastes with municipal solid waste and deposits them in landfills and dumps.

During the 1980s, urban population grew by 60% in Africa and by 25% in Asia and Latin America, and over 130 million of the world's poorest of the poor lived in cities (Panayotou, 1991, pp. 15–18). By the end of the 1990s, one-fourth of the total population of developing countries will be living in some 500 cities of more than 500,000 population, including 50 cities of over 4 million inhabitants (Bartone *et al.*, 1994).

During the period 1990–2030, forecasts predict that the world's population will grow by 3.7 billion people, with 90% of this growth in the developing countries and that same percentage of growth in urban areas (World Bank, 1992, pp. 7–8). Under these trends, economic output in developing countries will increase by 4–5% a year and be five times its present level by the year 2030 (World Bank, 1992, p. 9). The combination of rapid urban population growth and an expanding quantity of contaminated land in urban areas increases the potential to expose a large population to toxic chemicals.

CONDITIONS IN DEVELOPING COUNTRIES

The worst toxic pollution in the developing world comes from heavy metals from smelters and manufacturing plants, especially in Eastern Europe, and from chemical and fertilizer plants in Latin America, Asia, and Eastern Europe (World Bank, 1992, p. 19). Small-scale and cottage industries also make a substantial contribution to industrial production and pollution, although they typically do not produce as much toxic wastes as the larger industries. These small-scale industries include tanneries, textile-dyeing plants, dyestuff producers, metal working and electroplating shops, foundries, automobile repair shops and gas stations, battery production and recycling facilities, pesticide formulation and mixing facilities, paint manufacturers, and printing plants (Bartone and Benavides, 1993).

Even though the volume of toxic wastes being produced in the developing countries is increasing rapidly, it is still below the level found in the industrial economies. The latter countries produce 5000 tons of hazardous materials for every billion dollars of gross domestic product (World Bank, 1992, p. 54). The United States produces more than 200 million tons of hazardous wastes every year (Mazmanian and Morell, 1992), and Sweden produces about 500,000 tons per year (Lidskog, 1993).

Even so, the volumes of toxic substances that developing countries release to the environment each year is enormous. Malaysia produces 100,000 tons of hazardous waste per year (*The Economist,* 1994, p. 61). In 1969, Thailand had about 500 factories and roughly half of them produced hazardous wastes. There are now 26,000 factories producing hazardous wastes in Thailand and their number could triple in a decade (World Bank, 1992, p. 55). Hungary produces 5 million tons of hazardous wastes annually, with 3.2 million tons stored on site and 1.8 million tons disposed in lagoons or landfills (Reiniger, 1992). In São Paulo, Brazil, waste collection mixes industrial waste with domestic waste in the 16,600 tons of solid waste generated daily. Experts estimate that 20% of that industrial waste is hazardous (Bartone *et al.,* 1994, p. 27).

In one or two decades at present rates of industrial growth, many of today's developing countries will be producing quantities of hazardous wastes comparable to current levels in the developed countries. This is alarming, because developing countries have neither effective policies nor the physical infrastructure to safely handle hazardous materials and dispose of hazardous wastes. Open dumping of solid hazardous wastes and the disposal of liquid hazardous wastes into sewers, rivers, ponds, or the ground remain the primary disposal options in many developing countries.

CONTAMINATION AND ENVIRONMENTAL CLEANUP POLICIES IN DEVELOPING COUNTRIES

Some developing countries are further along in addressing their contamination and cleanup problems. Thailand is currently developing a set of policies, and though many of the projects are on a small scale, successful pilot projects may evolve into an effective package of policies. For example, in Bangkok, hundreds of small electroplating, galvanizing, and auto repair shops have contracts with a hazardous waste treatment facility constructed by the public sector and operated by the private sector (Panayotou, 1991, p. 91).

Hong Kong has serious toxic contamination problems but has started taking action to prevent additional contaminated areas. Many of the 50,000 small- to medium-sized enterprises release hazardous wastes onto the land or into local sewers (Panayotou, 1991, p. 83). In 1993, the U.S.-based Waste Management International opened a hazardous waste treatment plant in Hong Kong that was paid for by the government, which plans to charge user fees to industries. Studies show that large companies were likely to use the plant because of concern for their public image. Small companies, however, were likely to continue their standard practice of dumping wastes into the sewers (*The Economist,* 1994, p. 61). The Hong Kong government is considering a levy on chemicals imported into Hong Kong to help support the plant. There is a communal hazardous waste treatment facility at Tuen Mun in the New territories, Hong Kong, that serves three leather tanneries (Panayotou, 1991, p. 83).

Malaysia is also planning to develop its hazardous waste disposal infrastructure. It arranged for Kualiti Alam, a consortium of Malaysian companies, and Kruger, a Dutch waste treatment firm, to build a treatment plant at Bukit Nanas, 50 kilometers (31 miles) south of Kuala Lumpur. The plant will include an incinerator. The plan calls for the consortium to recover the cost of 400 million ringit (approximately $154 million U.S.) from charges for hazardous waste treatment with no government subsidies (*The Economist,* 1994, p. 61). Treatment cost estimates are higher than in Denmark, where costs are subsidized by the state. Plans estimate Malaysian incineration costs to be about $1000 per ton, roughly 50% higher than the European average (*The Economist,* 1994, p. 61). A hazardous waste treatment plant planned for Selangor, Malaysia, will include an incinerator, a chemical treatment unit, sludge stabilization, and a landfill. It will serve small-scale plants that would otherwise be unable to properly treat their hazardous wastes (Panayotou, 1991, p. 83).

The Philippines has 85,000 small manufacturing plants in major cities that discharge their wastes onto the land (Panayotou, 1991, p. 83). The Philippines Industrial Waste Exchange operates a clearinghouse for waste generators and users to promote the recycling of hazardous waste (Panayotou, 1991, p. 83).

Egypt uses communal toxic and hazardous waste processing facilities. In Cairo, the government encourages small-scale industries to relocate to satellite cities where the government provides communal facilities at nominal charges (Bartone and Benavides, 1993, p. 4). This policy both encourages treatment of hazardous wastes and removes waste-producing industries from locations in close contact with dense residential populations.

As part of a policy to prevent contaminated sites, some of the states in Mexico are also using hazardous waste processing facilities to induce industries to relocate. The city of Leon in Guanajuato State produces about 5.8 million hides per year from its tanneries (Bartone and Benavides, 1993, p. 8). The tanning process uses chromium, a toxic heavy metal. The tanneries in Leon release 483 tons of chromium per year, mostly to sewers or the Turbio River. The city is using a government subsidy of 70% of its operating cost to build an industrial park with treatment systems for tanneries (Bartone and Benavides, 1993, p. 9).

Hungary has also implemented the beginnings of a policy to prevent contamination. It now has a national waste disposal plan to reduce the production of hazardous wastes and promote their safe disposal. Early parts of the plan were a hazardous waste disposal site put in operation in Aszod in 1989 and a hazardous waste incinerator in Dorog started in 1990 (Reiniger, 1992).

Zimbabwe has no special national legislation for hazardous waste management. Consequently, the chemical industry produces and buries large volumes of hazardous wastes in the ground (Bartone and Benavides, 1993, p. 12). However, Zimbabwe does have some discrete programs designed to lessen the contamination and environmental cleanup problem. For example, the City of Harare Department of Public Works operates an oil recycling program that recovers 100,000 liters of oily wastes per month from garages (Bartone and Benavides, 1993, p. 13).

CONCLUSIONS

Developing countries must take action specifically designed to contend with their contamination and environmental cleanup problem. Policies for economic growth can complement those for environmental protection, however, enlightened economic and environmental policies designed to provide market incentives so firms do not pollute are not enough to prevent the problem from becoming worse.

Policies concerning hazardous wastes require a comprehensive set of standards and regulations to manage hazardous materials and wastes from their point of generation to their final disposal site. National circumstances will determine the policies and specific instruments needed by developing countries. Administrative capabilities are often a critical factor in environmental policies, as many studies report the inadequacy of existing institutions and personnel to carry out effective monitoring and enforcement activities (Bernstein, 1993, p. 26). For this reason, blunt environmental policy instruments or policies that establish self-enforcing incentives may be attractive in developing countries.

Such policy instruments indirectly influence potentially polluting activities. An example is establishing a tax on imputs that generate toxic wastes after industrial processing. Inputs could include chemicals and even specific technologies. Whereas a direct policy instrument would regulate the generation of toxic waste, the blunt instrument taxes the inputs, which only indirectly provides an incentive to reduce the volume of toxic waste.

The advantage of blunt policy instruments is administrative. Instead of requiring a sophisticated regulatory agency to monitor and inspect firms, the blunt instrument can use the existing tax collection system to provide an incentive to minimize the use of toxic-waste-producing chemicals. In situations where thousands of small firms are using hazardous wastes, the monitoring and enforcing of regulations require a large and highly trained staff. In many developing countries, government demands make administrative expertise a scarce commodity. Thus blunt policy instruments have inherent advantages in preventing pollution.

Developing countries should carefully marshal their administrative capabilities toward preventing new contaminated sites, and the sources of toxic pollutants posing the greatest human health risk should receive the highest priority. If developing countries hope to successfully contend with contamination and cleanup, they must start immediately.

References

Bartone, C. and L. Benavides. 1993. Local management of hazardous wastes from small-scale and cottage industries. Presentation at the Fifth Pacific Basin Conference on Hazardous Waste, East-West Center, Honolulu, November.

Bartone, C., J. Bernstein, J. Leitmann, and J. Eigen. 1994. *Toward Environmental Strategies for Cities: Policy Considerations for Environmental Management in Developing Countries.* Urban Management Programme Policy Paper 18. Washington, DC: World Bank.

Bernstein, J. 1993. *Alternative Approaches to Pollution Control and Waste Management: Regulatory and Economic Instruments.* UNDP/UNCHS/World Bank Urban Management Programme Discussion Paper No. 3, p. 26. Washington, DC: World Bank.

The Economist. 1994. A sniff of the East. Vol. **332**(7870), 61.

Forrest, A. 1995. A turning point? *Environ. Forum* **12**(2), 24–30.

Lidskog, P. 1993. Whose environment? Which perspective? A critical approach to hazardous waste management in Sweden. *Environ. Plann. A* **25**(4), 571–588.

Lucas, R. 1992. *Toxic Releases by Manufacturing: World Patterns and Trade Policies,* Working Paper WPS 964. Washington, DC: World Bank.

Lucas, R., D. Wheeler, and H. Hettige. 1992. *Economic Development, Environmental Regulation and the International Migration of Toxic Industrial Pollution: 1960–1988,* Background Paper No. 33 for World Development Report 1992. Washington, DC: World Bank.

Mazmanian, D., and D. Morell. 1992. Beyond Superfailure: America's Toxics Policy for the 1990s. Boulder, CO: Westview Press Inc.

Organization for Economic Cooperation and Development. (OECD) 1991. *Draft Guiding Principles for Chemical Accident Prevention, Preparedness and Response.* Paris: OECD, Environment Directorate, Environment Committee.

Panayotou, T. 1991. *Managing Emissions and Wastes,* Background Paper No. 36 for World Development Report 1992. Washington, DC: World Bank.

Reiniger, R. 1992. Overview of Hungary's environmental situation. *Summary Report.*

The 1992 NATO/CCMS Pilot Study Meeting on Evaluation of Demonstrated and Emerging Technologies for the Treatment and Clean-up of Contaminated Land and Groundwater, Budapest, October.

United Nations Environment Programme. 1988. *APELL Awareness and Preparedness for Emergencies at Local Level: A Process for Responding to Technological Accidents.* Paris: UNEP Industry and Environment Office.

United National Environment Programme. 1992. *Hazard Identification and Evaluation in a Local Community.* Industry and Environment/Programme Activity Center (IE/PAC), Technical Report No. 12. Paris: UNEP.

World Bank. 1985. *Environment, Health and Safety Guidelines for Use of Hazardous Materials in Small and Medium-Scale Industries.* Office of Environmental and Scientific Affairs. Washington, DC: World Bank.

World Bank. 1992. *World Development Report: Development and the Environment.* New York: Oxford University Press.

CHAPTER 8

THE SUPERFUND PROGRAM IN THE UNITED STATES

INTRODUCTION

The U.S. Congress created the Comprehensive Environmental Response, Compensation, and Liability Act (CERCLA) of 1980 to clean up the country's *most dangerous* contaminated sites. The basic idea of the Superfund was to establish a "shovels first, lawyers later" program to enable fast action to protect the public and to complete environmental cleanups of contaminated sites. EPA relaxed this policy somewhat after criticism that they were spending too much trust fund money and being too easy on the parties responsible for the contaminated areas. EPA still attempts to clean up the sites quickly, but now uses an "enforcement first" strategy. This strategy attempts to increase the number and dollar value of private sector "potentially responsible party" (PRP) settlements and to maximize the leverage of trust fund money (Probst and Portney, 1992, p. 21).

CERCLA established an initial trust fund of $1.6 billion from several sources, including special taxes and general revenues. Congress gave EPA powerful enforcement authority to make responsible parties either clean up the site themselves or reimburse EPA for government-funded cleanups.

The Superfund program has achieved substantial success. It has remediated many contaminated sites, protected public health, and improved the quality of life of Americans in many cities, towns, and neighborhoods. It has stimulated the development of new and improved remediation techniques and a new industry to investigate and conduct environmental cleanups. It has dramatically raised awareness of toxic substances in firms producing and using them and by the insurance, banking, and real estate industries.

Despite these substantial accomplishments, there is widespread support to change the Superfund program. The rest of this chapter examines various criticisms of the program with the goal of discovering how the United States can reform the program while retaining or increasing its environmental, public health, and social achievements.

TWO TYPES OF SUPERFUND ACTIONS

The Superfund program includes two types of actions: removal actions and remedial actions. Removal actions are short-term emergency actions intended to protect human health and the environment from serious immediate threats.

An example of a removal action is the Robson residence in Maine (U.S. Environmental Protection Agency, 1993a), where government officials discovered chemicals when a widow asked advice about the chemical laboratory her husband left in the basement. EPA removed 2000 containers of chemicals, including explosive, hazardous, and radioactive compounds from the site. The EPA may use funds from the CERCLA trust fund to respond to hazardous waste emergencies at both National Priorities List (NPL) and non-NPL sites. The EPA had completed more than 3200 emergency actions at 2540 sites by 1993 (U.S. General Accounting Office, 1993a). Most participants and observers consider removal actions under the Superfund program to be successful and they are not subject to the intense criticism that remedial actions receive. Some proposals for reforming Superfund include greater use of removal actions (U.S. General Accounting Office, 1996).

Remedial actions are longer-term, comprehensive, and restorative environmental cleanups of sites that EPA places on the National Priorities List. An example of a remedial action under the Superfund program is the Triana/Tennessee River site in Alabama, which is within the boundaries of a national wildlife refuge and a U.S. Army installation (U.S. Environmental Protection Agency, 1993b). Industrial production and disposal of DDT on the site began in 1947 and continued until the plant closed in 1970. Approximately 409 tons of DDT and the related breakdown products DDD and DDE contaminated the sediments of 11 miles of two tributaries of the Tennessee River. The contamination threatened 600 residents of nearby towns, bald eagles, 60,000 migratory birds, and the habitat of several endangered species.

Government environmental officials started actions at the Triana/Tennessee site in 1977 when they issued a fish advisory. In 1981, EPA proposed the site for the NPL; in 1983, it listed the site. The EPA accepted

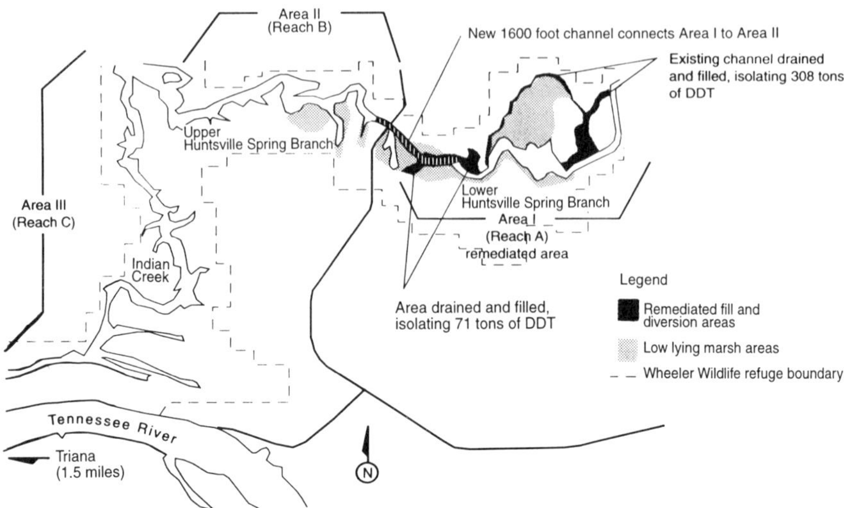

FIGURE 8.1 The EPA and Olin cleanup site on the Triana/Tennessee River, Alabama. [From U.S. Environmental Protection Agency (1993b).]

the cleanup plan for the site in 1984, and remediation activities began in 1986. Remedial actions included diverting the river to a new channel and containment in-place of the contaminated sediments in the old river channel (Fig. 8.1), as well as removing 150,000 cubic yards of contaminated soil and adding 217,000 cubic yards of clean soil and 190,000 cubic yards of rock to fill in the old river channel. These remedial actions encapsulated 93% of the contaminated sediments in the river system. In 1991, cleanup activities were completed, although monitoring will continue until 1998. EPA estimates the cost of the environmental cleanup at the Triana/Tennessee River site to be $30 million.

Some contaminated sites receive both removal and remedial actions. For example, EPA performed a removal action at the Valley of the Drums site in Bullitt County, Kentucky, in 1981, and after adding the site to the NPL in 1983 it began remedial actions in 1986 (U.S. Environmental Protection Agency, 1992a).

HISTORY OF THE SUPERFUND PROGRAM

The U.S. Congress arbitrarily established the number of initial sites on the NPL at about 400 to approximate the number of congressional districts in the United States. When the EPA announced the initial list of Superfund sites in 1983, there were 406 contaminated sites on the NPL. The NPL had grown to 1275 sites by 1993, and EPA anticipates close to 2000 sites by the year 2000 (U.S. General Accounting Office, 1993a). Of the 1275 sites on the NPL as of September 1993, there were 56 sites with action not begun, 696 sites in study or design stage, 374 sites in cleanup, 109 sites with construction complete, and 40 sites deleted from the NPL (U.S. General Accounting Office, 1993b). The pace of completions has improved recently. As of September 1995, construction was complete at 304 sites and EPA had deleted 84 sites from the NPL (U.S. General Accounting Office, 1996).

The Superfund program has far exceeded expectations about its size and cost. In addition to cleanup expenses by private responsible parties, EPA spends money from the Superfund program's trust fund on environmental cleanup. Superfund has been reauthorized twice with cumulative funding of $15.2 billion. In 1994, the U.S. Congress considered reauthorization but took no action. The $15.2 billion now authorized likely will be insufficient to cover EPA's cost to clean up those sites already on the NPL. EPA's estimated share of the cost of cleaning up those sites is $27.2 billion (U.S. General Accounting Office, 1993b).

In 1994, the Congress came close to amending and reauthorizing the Superfund program but failed when the congressional elections approached. The Clinton administration's proposals had considerable support for two important changes in the program: (1) replacing protracted litigation with arbitration and (2) spreading the cleanup costs among all the companies responsible for polluting a site, in place of current law that can hold a single party liable. At the last minute, Congress failed to reauthorize the Superfund program and attempts since then to reauthorize it have been unsuccessful.

Other components of the failed 1994 proposed amendments were interesting. One included a uniform national standard, to be negotiated after Congress reauthorizes Superfund, that would determine the allowable degree of health risk at any site after the completed environmental cleanup. These standards vary widely now. Other proposals would allow flexibility on how thoroughly the remediation effort should clean a site. EPA would base the extent of remediation on considerations such as the future use of the land. The proposed amendments also included new arbitration procedures and a requirement to consider cleanup costs when deciding how thoroughly to clean a site. Though there was not agreement, there was also considerable support for dropping the current law's retroactive liability provisions that can hold companies responsible for cleaning up waste that it dumped legally before 1980. In 1995, EPA initiated a series of administrative reforms that implemented some of the reforms that had been part of the proposed modifications in the failed 1994 congressional reauthorization effort (Herman, 1995).

THE SUPERFUND PROCESS

The Superfund system is complex. Once the government identifies a site that may require a cleanup, officials place the site on an inventory for further investigation. For example, in 1984, local officials investigated reports of alarming numbers of birds dying on the Tri-State Plating site in Columbus, Indiana. They found that the deaths resulted from drinking water from contaminated puddles (U.S. Environmental Protection Agency, 1993c). This incident contributed to EPA investigations of this electroplating facility in 1984, which was located in a residential area and threatened the health of 30,000 people. EPA placed the site on the NPL in 1985.

At any time, a Superfund site may receive a Removal Action. Emergency conditions may require fast action or the site could worsen before EPA could complete a remedial cleanup. At the Tri-State site, a removal action took 27 drums of electroplating waste from the abandoned facility to a licensed hazardous waste facility. The remediation team also constructed a fence around the Tri-State site to prevent public access.

In the preremedial process, sites receive a Preliminary Assessment (PA), and some then go forward to a Site Inspection (SI), with some of those sites scored by the Hazard Ranking System (Fig. 8.2). EPA uses the Hazard Ranking System to determine which sites are so seriously contaminated that EPA should put them on the NPL for remediation under the Superfund program. The NPL includes only those sites scoring 28.5 or higher out of a possible 100 points.

Evaluating a contaminated site using the Hazard Ranking System costs an average of $176,000 (U.S. Environmental Protection Agency, 1990a). The EPA has revised this system several times, including expanding the number of exposure pathways considered, modifying the cutoff point used to determine NPL listing, and more heavily weighting ecological risks (U.S. Environmental Protection Agency, 1990a). As of March 1995, EPA had 15,723 sites identified for evaluation, with 1363 sites on the NPL (U.S. General Accounting Office, 1995a).

If the Hazard Ranking System score is high enough, EPA lists the site on the NPL. Listed sites then become eligible for a remedial cleanup paid for by responsible parties identified as having contributed to creating the uncontrolled toxic waste site or by the government, if necessary. Under current procedures, EPA lists on the NPL only about 10% of sites proposed for the Superfund program.

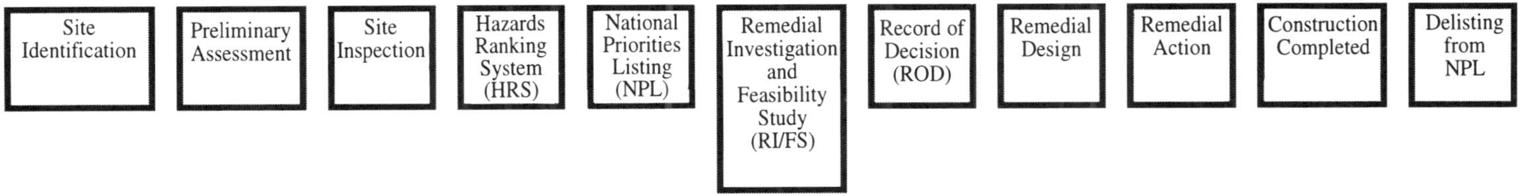

FIGURE 8.2 Steps in the process of the U.S. Superfund program.

Some states have their own mini-Superfund programs with lists of sites that require cleanup and that are not on the NPL. Generally, the most seriously contaminated sites become Superfund sites and state programs take the responsibility of cleaning up the less seriously contaminated sites; however, this is not always the situation. Researchers estimate that state cleanup programs will remediate 5 to 15 times as many contaminated sites as the federal Superfund program (Hird, 1994, p. 22).

The Hazard Ranking System is controversial. The score produced by it is one way to distinguish less serious from more serious contamination. However, researchers suggest that only a tenuous relationship exists between the score and the site's hazard potential, and indeed, the cutoff point of 28.5 is arbitrary. Researchers estimate that as many as 2000 or more out of the 17,000 sites rejected for the NPL may be false negatives (U.S. Congress, 1989, p. 114). This suggests that cost-effective remediation of those sites is possible, but EPA has not listed the site on the NPL. Most damning of all the criticisms of the Hazard Ranking System is that there are thousands of possible sites not on the NPL because there is no systematic program to search for the most serious contaminated sites in the United States.

NPL sites receive a Remedial Investigation and Feasibility Study (RI/FS) to define contamination and environmental problems and to evaluate cleanup alternatives. In May 1995, EPA issued a directive that remedial action objectives developed during the RI/FS should reflect the reasonably anticipated future land use or uses (U.S. Environmental Protection Agency, 1995). This is a potentially important reform of the Superfund program. This approach may limit remedial actions to allow levels of residual contamination that the government considers safe for the use of the restored land. For some contaminated sites, less expensive remedial actions can be used because higher levels of residual contamination after remediation are acceptable. Under this approach, the remediation of contaminated sites that will be used for an industrial warehouse might not have to be as thorough as that required for land that will be the site of a school or residence (this issue is explored more thoroughly in Chapter 11).

The RI/FS stage of the Superfund process gives the public an opportunity to comment on the RI/FS and on EPA's preferred cleanup alternative. For example, at the Tri-State Plating site, EPA held extensive open meetings, published a regular site fact sheet, and set up a public education program because of the great public concern resulting from the site's location in the heart of a residential neighborhood (U.S. Environmental Protection Agency, 1993c).

After the RI/FS, EPA issues a Record of Decision (ROD) that records what remedy the government has chosen and the reasons for doing so. The decision may be that no cleanup is necessary. A ROD may only deal with part of a site's cleanup and several RODs may be necessary for a site. The ROD also contains a summary of EPA's responses to public comments. EPA chooses the site-specific cleanup goals and remediation technology in the ROD. In practice, EPA often selects multiple technologies at many Superfund sites.

The ROD is like a contract in which the government makes a commitment to actions that will constitute the environmental cleanup. The cleanup objective is to make the site safe, which means that after cleanup the site will pose only a minimal level of risk to human health and the environment. The government, people living nearby, and responsible parties all participate in defining the ROD.

If responsible parties agree to clean up the site, they sign a negotiated consent decree with the government. A consent decree stipulates the exact details of how the responsible parties will proceed. At the Tri-State Plating site, the company declared bankruptcy and EPA used the Superfund program's trust fund to pay for the environmental cleanup (U.S. Environmental Protection Agency, 1993c). If the cleanup uses Superfund money, the cleanup proceeds and EPA takes legal action that attempts to force the responsible parties to reimburse the Superfund for the cost of cleanup. When EPA uses Superfund money, the state government must agree to pay 10% of the cleanup cost.

In the post-ROD process, the site receives a Remedial Design (RD) study that specifies the engineering and construction details. The environmental cleanup begins with Remedial Action (RA), the actual implementation of the selected remedy. The remedial actions at the Tri-State Plating site included removal of the buildings, excavation and removal of contaminated soil from the entire site, spreading of clean soil, and planting grass on the site (U.S. Environmental Protection Agency, 1993c).

The stage known as Construction Completed indicates that the project team has completed all construction necessary for the remediation, but that remediation is ongoing. Many cleanups include long-term monitoring to determine whether the cleanup is effective and if additional cleanup is necessary. EPA may reopen and amend a ROD if the project team encounters new information or difficulties during the design and remedial action. When the project team considers a cleanup complete and effective, the EPA can "delist" a site from the NPL.

FINANCING THE SUPERFUND PROGRAM

The Superfund is really a trust fund financed from a variety of sources. In 1980 when the Superfund program started, the U.S. Congress financed the original $1.6 billion trust fund principally by taxing petroleum products and 42 chemicals. The logic of these taxes was that these industries produced the bulk of hazardous wastes in the past and therefore were linked to the contaminated sites. Thus, the original financing of the trust fund was tenuously related to the polluter pays principle.

When Congress reauthorized the Superfund program in 1986, the trust fund needed substantial enlargement. EPA needed a larger pot of money in the trust fund for several reasons. First, the extent of contamination was greater than anticipated when Congress passed CERCLA in 1980. Second, more money for cleanup than anticipated was being drawn from the trust fund, as opposed to coming from responsible parties. These parties were also taking longer than expected to reimburse the trust fund, and a large number of cleanups received no reimbursement.

During the debate over reauthorization in 1986, there was wide recognition that the trust fund needed additional sources of revenue. Many felt that the international competitiveness of the U.S. chemical industry would suffer if they significantly increased the size of the trust fund without enlarging the tax base used to supply it. The government proposed taxes on those industries that historically had contributed most to contamination, but reached a compromise that based financing on a combination of revenue sources. The total of $8.5 billion needed to enlarge the Superfund program over five years would now come from the following sources:

- $2.75 billion from a revised tax on petroleum products;
- $2.5 billion from a 0.12% tax on minimum corporate income over $2 million;
- $1.4 billion from a revised chemical feedstocks tax;
- $1.25 billion from general revenues;
- $0.6 billion from recovered costs from liable parties and interest on unused portions of the trust fund. (Hird, 1994, p. 126)

The Superfund program adheres to the polluter pays principle more in theory than in practice. The political rhetoric over funding cleanups suggests that responsible parties contribute all the costs. The actual amounts received from application of the polluter pays principle are much less. Through the end of fiscal year 1994, EPA had spent over $10.1 billion on environmental cleanup of Superfund sites but had recovered less than $1 billion from potentially responsible parties (PRPs) (U.S. General Accounting Office, 1995a). The remaining $8.7 billion includes unrecoverable costs and costs that EPA is currently attempting to recover through litigation.

Supporters of the Superfund program in the United States used the substantial political popularity of the polluter pays principle to enact the CERCLA legislation in 1980 and enlarge the program in the following years. As the funding sources proposed in the 1986 reauthorization of

Superfund indicate, the direct application of the polluter pays principle is small. Only $0.6 billion of the total $8.5 billion (7%) came directly from cost recoveries from responsible parties, with some of that amount from interest on the fund. The overall operation of the Superfund program stretches the polluter pays logic considerably. It is difficult to claim that taxing industries, which are operating legally today, because of a historical connection to activities that caused some of the contamination, is an application of the polluter pays principle. However, if we accept this contorted logic, the recovered $0.6 billion plus the $2.75 billion from the tax on petroleum products and the $1.4 billion from the revised chemical feedstocks tax, all totaling $4.75 billion, would fall into this category. This amount is slightly more than half (56%) of the entire $8.5 billion trust fund. The $2.5 billion from a 0.12% tax on minimum corporate income over $2 million and the $1.25 billion from general revenues clearly do not implement the polluter pays principle.

PROBLEMS OF SUPERFUND

The U.S. Superfund program is obviously a large and expensive program that is highly controversial. Some argue that it is more "pork barrel" than a serious environmental program, as indicated by the original intent to spread spending across congressional districts (Crandall, 1988). Other analysts conclude that Superfund is not a "pork barrel" program (Hird, 1994, p. 160). One critical report concludes:

> The fundamental problems of Superfund are that the magnitude of the problem is still unknown, the program is in several aspects both inefficient and inequitable, the public is alienated from the process and does not trust the perspectives that EPA and risk experts bring to the issue, and finally that the program continues to experience substantial political and public relations problems. (U.S. Congress, 1988, p. 182)

The EPA has made significant improvements in the Superfund program since that harsh evaluation in 1988 (Romano and Simon, 1993). Yet despite an improved program, the 1988 criticisms remain pertinent.

A review of these criticisms is helpful in understanding the issue of contamination and policy to effect the cleanup of contaminated sites. In the following sections, we review the problems with Superfund in the following areas: results, management, public perception, and cost.

Results Problems

One of the most serious problems with the Superfund program is that NPL sites represent little real risk. Although some sites on the NPL are a serious threat to public health and the environment, others are not. Some NPL sites do not warrant the elaborate procedures and high expense of cleanup. The actual health risks from some sites are small compared to many other environmental problems, and especially small when compared to other social problems (U.S. Environmental Protection Agency, 1987, 1990d). Though there are many confirmed health risks in our environment that we do not attempt to reduce, the Superfund program spends large amounts of funds on *potential* health risks.

For example, the EPA included the Army Creek Landfill site near New Castle, Delaware, on the original NPL in 1983 because of the *potential* contamination of groundwater supplies of drinking water (U.S. Environmental Protection Agency, 1993d). Yet no contamination ever reached a public drinking water supply well to cause a health risk. In 1973, local government had installed a series of groundwater recovery wells that effectively prevented the contamination from reaching public supply wells. However, there was the potential that the contamination might someday threaten 5000 people, which resulted in a high score on the Hazard Ranking System and the expenditure of more than $26 million to complete an environmental cleanup.

In general, health risks are largely unknown and certainly unproved. There is a critical absence of information on the distribution of exposures and the health effects associated with contaminated sites (National Research Council, 1991). The best evidence we have about the actual risks of such sites suggests that the real risks are not as large as many other risks that receive far fewer resources (Portney, 1988; Hird, 1994, p. 91).

On the other hand, some sites are not on the NPL but should be. The U.S. government has no coordinated site discovery process that attempts to identify the most seriously contaminated sites in the country for inclusion in the Superfund program. The most serious health risks usually result from groundwater contamination that may go undetected for years. Many Superfund sites are serendipitously discovered. Officials investigated the Tri-State Plating site because of reports of dying birds (U.S. Environmental Protection Agency, 1993c). Government officials discovered dangerous chemicals left in the basement of the Robson residence for 18 years only when someone asked a knowledgeable person for advice (U.S. Environmental Protection Agency, 1993a).

Since there is no organized attempt to identify the most contaminated sites in the United States, we will continue to find existing but unidentified sites. Research suggests that newly identified contaminated sites are just as hazardous as sites already on the NPL. The implication is that after 16 years of Superfund, we have not yet found the most serious contaminated sites.

The process to prioritize cleanup for NPL sites is arbitrary and bears little relationship to the actual risk that sites represent (U.S. Environmental Protection Agency, 1990b). There are no general principles to determine "how clean is clean" in setting standards for site remediation. Thus the Superfund program "gold plates" some sites while not providing other sites with any serious remediation (Hird, 1994, p. 113). The government might better spend some of the funds expended on "gold-plated" cleanups to solving environmental or public health problems related to air pollution, radon exposure, or other problems unrelated to contaminated sites.

There is also confusion over the Superfund program's goals. The goal of remediating the worst sites first is different from the goal of eliminating the greatest public health and environmental risk most efficiently. Nor does Superfund consider the cost of remediation and the level of risk reduction (Hird, 1994, p. 191). Policy could try to maximize the risk reduced per dollar spent: aim at those contaminated sites where the ratio of risk averted per dollar spent is highest. This approach might allow the prevention of more future cancers and illnesses by shifting money to more efficient public health and environmental programs (National Research Council, 1991).

Another criticism is that cleanups at Superfund sites take too long. Though there is considerable variation in the duration of cleanup, the average Superfund cleanup takes from 11 to 15 years from initial identification until the project team implements the remedy (U.S. General Accounting Office, 1991; Probst and Portney, 1992, p. 20). Actual completion of the remediation takes longer.

Even so, some cleanups at Superfund sites will not be effective in the long term. The selection of remediation techniques, especially containment, suggests that about 75% of Superfund cleanups are unlikely to last over long periods (U.S. Congress, 1989, p. 12). This criticism applied especially to the early years of the Superfund program. After 1986, EPA put a greater emphasis on destruction of contaminants rather than containment. Recent proposals suggest a return to an increased emphasis on containment (U.S. General Accounting Office, 1996). Table 8.1 summarizes the problems with the Superfund program.

Management Problems

There is broad agreement that the EPA has greatly improved Superfund management in recent years; there is also agreement that considerable room still exists for improvement (U.S. General Accounting Office, 1995b). Table 8.2 reviews the management problems of the Superfund program.

To begin with, the relatively young cleanup work force needs better

TABLE 8.1

RESULTS PROBLEMS WITH THE SUPERFUND PROGRAM

1. Some sites on the NPL represent little real risk and should not be remediated.
2. Some sites are not on the NPL but should be.
3. There are no general principles to determine "how clean is clean?" Thus some sites are "gold plated" while others do not receive any serious remediation.
4. There is confusion over the goals of the program.
5. Some environmental cleanups at Superfund sites will not be effective over the long term, particularly containment operations.

management, information, and technical assistance (U.S. Congress, 1989). Environmental cleanup of contaminated sites is a relatively new activity and would benefit from a better-trained and experienced work force and more experienced managers.

At present, management procedures favor remediation techniques with a proven track record, despite the availability of new and potentially more effective and less costly remediation techniques (U.S. Congress, 1989, p. 110; Hird, 1994, p. 113). "As long as the EPA managers at toxic waste sites are rewarded for how many sites they clean up quickly, they are never going to select methods like bioremediation" (Hird, 1994, p. 30). Though this comment is generally accurate, Superfund does encourage new and innovative remediation technologies through funding incentives, and some Superfund sites do use bioremediation (see Chapter 2), for example, the French Limited site in Harris County, Texas, and the Brown

TABLE 8.2

MANAGEMENT PROBLEMS WITH THE SUPERFUND PROGRAM

1. The relatively young cleanup workforce needs better management, information, and technical assistance.
2. Management procedures favor remediation techniques with a proven track record despite the availability of new and potentially more effective and less costly remediation techniques.
3. Management procedures favor completion of cleanup projects rather then the environmental cleanup of the contaminated sites that pose the greatest risk.
4. The EPA has not managed the Superfund program in a manner that produces similar environmental cleanups at similar contaminated land sites.

Wood Preserving site in Live Oak, Florida (U.S. Environmental Protection Agency, 1993e,f).

Management procedures also favor rapid completion of cleanup projects rather than the cleanup of sites that pose the greatest risk. Speed of cleanup is important to EPA to counter congressional criticism of delays in the program. Why should EPA tackle a particularly difficult site with groundwater pollution when management receives the same credit for rapid and permanent cleanup of an uncomplicated site? EPA's management focus on the number of sites cleaned produces a "bean counter" approach to measuring the success of the Superfund program (Hird, 1994, p. 32).

Nor has the EPA managed Superfund in a manner that produces similar cleanups at similar contaminated sites. Some critics think this is inequitable. They argue that EPA's concerns about controlling expenditures from the program's trust fund result in less thorough cleanup of some sites and excessive negotiation with PRPs. Under these conditions, the PRPs may do a less expensive and thorough cleanup than a government-led cleanup funded from the trust fund (U.S. General Accounting Office, 1992). In 1988, researchers found that 75% of PRP-led cleanups resulted in containment of the contamination, whereas 78% of government-led cleanups resulted in waste destruction (U.S. Congress, 1989, p. 6).

Public Perception Problems

All parties involved in the Superfund program have their own perspectives about the program. By not setting clear priorities for Superfund, Congress has fed unrealistic expectations that produce many perceived problems. Some of the issues that specific parties perceive as problems result from the unique perspective of that party.

For example, local communities that discover a contaminated site that becomes a Superfund site often perceive the process differently from the other participants. When EPA places great reliance on risk assessments, cost–benefit analyses, and other technical studies, communities do not believe them (U.S. Congress, 1989, p. 9). Local governments feel that they are unable to effectively influence the process and as a result that they are subject to seemingly endless delays in cleanup. During the 1994 failed Superfund reauthorization efforts, a House subcommittee voted unanimously to approve a bill that would have increased the influence of local communities (*Environmental Damage Valuation,* 1994). Table 8.3 presents a summary of such perceived problems, though in some cases they are based on a party's self-interested perspective and may not be seen as problems by other parties.

TABLE 8.3

PERCEIVED PROBLEMS WITH THE SUPERFUND PROGRAM[a]

1. Affected individuals and communities feel left out of the Superfund process.
2. Environmentalists think EPA takes far too long to remediate sites, that cleanup standards are too lax and inconsistent from site to site, and that the fund is too small to meet cleanup needs.
3. Chemical and petroleum companies believe that they are unfairly singled out in funding the program through excise taxes on their products, in addition to being major PRPs who are called upon to finance cleanups.
4. PRPs complain of liability schemes that they believe are fundamentally unfair.
5. Municipalities feel they are unfairly dragged into the legal process when they had little part in causing the contamination, and that it was not the intent of Congress that municipalities be held liable for the cost of environmental cleanup.
6. Insurance companies feel that they are ensnared in a web of litigation and potential liability that limits their ability to underwrite insurance and that ultimately may threaten their financial viability.
7. Legislators feel frustrated because they want action to change the Superfund program, yet there are so many interest groups with conflicting positions that revision of the program is difficult.

[a]Many of these issues are modified from Hird (1994, p. 181).

Cost Problems

Controversies over the cost of the Superfund program include concerns about its magnitude, the equity of cost allocation, and its cost-effectiveness. Many of these arguments are powerful criticisms, however, by focusing on costs, they consider only easily quantifiable measures. But one should also consider the benefits that are difficult to quantify, such as deterrence of activities that may produce future contaminated sites, quality of life benefits, and the psychological benefits of cleaning up derelict eyesore sites in neighborhoods.

As discussed earlier, Superfund has grown much larger than the U.S. Congress originally intended, with NPL sites increasing from 406 to about 1300 between 1980 and 1995. The program continues to grow because sites are being added to the NPL at a faster rate than they are being delisted. It is likely that the NPL will soon include more than 2000 sites. Analysts estimate that implementing a program to actively evaluate contaminated sites could yield 10,000 NPL sites by the year 2000 (U.S. Congress, 1989, p. 11).

The average cost of a Superfund site cleanup has increased to more than $30 million. The less complicated and less contaminated sites were the first completed remediation sites delisted from the NPL. Thus the average cost of a Superfund cleanup will increase further as project teams complete more of the complex and more costly cleanups. For example, the cumulative estimated dollar value or responsible party work commitments for the cleanup of the Well G&H site in Woburn, Massachusetts, is $7.4 billion (U.S. Environmental Protection Agency, 1992e). This site contains two former municipal landfills. No one can predict the total cost of the Superfund program, but reputable predictions range from $750 billion to $1.7 trillion over the next 30 years (Russel *et al.*, 1991).

Despite the intention that Superfund clean up the worst sites, some of the most contaminated sites are remediated under state-operated programs or without any government program. For example, in Broward County, Florida, there are presently 979 identified contaminated sites. Yet only 6 (less than 1%) of those sites are on the NPL (Department of Natural Resources Protection, 1994). Although the great majority of these are sites where petroleum products have leaked, in some cases the health risks of these non-Superfund sites are as serious as those of some Superfund sites. Private firms also pay for the remediation of many contaminated sites, as they sometimes prefer to bear the entire cost of cleanup rather than become enmeshed in the costs, delays, and frustrations of the Superfund program (U.S. Environmental Protection Agency, 1990c). We must consider all programs that contribute to the cleanup of toxic and radiation contamination when we evaluate the cost of cleanup policies.

Some analysts believe that the cost of cleanups may be a serious economic burden, and that by diverting substantial resources away from other economic activities, these environmental policies may harm U.S. international economic competitiveness; this harm is known as "opportunity cost." Superfund spending may also make the cost of American products more expensive, at least in the short term. On the other hand, the costs of the Superfund program may be so small in relation to the overall economy that they have an insignificant effect on issues of international competitiveness. It is also possible that the long-term benefits of a cleaner environment outweigh the short-term costs of environmental cleanup.

Some critics claim that Superfund's cost is high because the program is inefficient, and because about 50% of Superfund cleanups address speculative future risks (U.S. Congress, 1989, p. 12). Sites characterized as representing only a speculative risk include two-thirds of all sites with groundwater contamination and one-third of sites with soil contamination. Expending limited funds on speculative risks preempts spending to identify and reduce real risks at other contaminated sites. An example of this is the Army Creek Landfill site near New Castle, Delaware, which was discussed in an earlier section (U.S. Environmental Protection Agency, 1993d).

The Superfund program's cost is also high because of the unnecessarily high or avoidable administrative and transaction costs. Legal fees for negotiation and litigation are the largest part of transaction costs. Researchers who examined environmental insurance estimate that 21% of total Superfund costs for sites with large numbers of PRPs go to transaction costs, and that 88% of environmental insurance payments go to transaction costs (Acton and Dixson 1992). Hazardous wastes disposal sites often have many PRPs. For instance, the Coal Creek site near Chehalis, Washington, is the location of an inactive transformer salvage facility where 88 parties agreed to perform the cleanup (U.S. Environmental Protection Agency, 1993g).

One way to evaluate the cost of an environmental policy is to use a statistic known as the "cost per statistical life saved" (CPSLS). CPSLS allows comparison of how much each policy would spend on average to save an additional human life. If we evaluate the highest-risk site from among all sites on the NPL, the expected number of people contracting cancer over their lifetimes would be between 3.25 and 73.52. If cleanup costs an average of $25 million per site, then the CPSLS is between $340,000 and $7.7 million. This is well within what individuals reveal themselves willing to pay to avoid a future statistical death and is consistent with other regulatory costs (Hird, 1994, p. 95). However, the NPL also contains sites that are not such serious risks. Using data from Superfund sites located in New England, the median risk of contracting cancer from a site was 1.3×10^{-5}. This corresponds to a CPSLS of $340 million to $7.7 billion. This is far less cost-effective than virtually any U.S. regulation, environmental or otherwise (Hird, 1994, p. 95).

There are often continuing costs at sites after remediation and a system to ensure the availability of funds for these costs is not yet in place. At almost half of the sites that have completed construction and been deleted from the NPL, there remain significant costs for oversight, operation, and maintenance. These costs are greatest where the cleanup entailed on-site containment. Responsible parties or the states must continue ongoing operations and maintenance after EPA delists the site, even when the EPA paid for the cleanup from the trust fund. An example is the Army Creek Landfill site, where the remediation team installed a protective cap to cover an abandoned landfill. This site requires long-term maintenance and monitoring (U.S. Environmental Protection Agency, 1993d). The EPA has already incurred costs of $374 million and estimates that states will incur costs of $1 billion between 1993 and 2000 (U.S. General Accounting Office, 1993b).

FAIRNESS OF THE SUPERFUND PROGRAM

The fairness of the Superfund program has always been a politically charged issue, beginning with the debate and passage of the original legislation (CERCLA) in 1980 (U.S. Senate, 1983, p. 153). The program has enjoyed widespread public support because the public perceives it as an effective response to an unfair situation: innocent citizens unknowingly exposed to the "ticking time bombs" of contaminated sites. Despite its intent to be fair, critics claim that certain aspects of the program, especially how it allocates the costs of environmental cleanup, are blatantly not fair.

When pollution occurs that produces a contaminated site, some people benefit from not paying the cost of disposing hazardous wastes in ways that we now consider appropriate. Many legal and accepted disposal practices of the past are now illegal. The Superfund program applies retroactive liability to these past acts, which makes parties liable for cleanup costs for acts that were legal at the time they occurred, but which caused contamination.

An example of retroactive liability is the Ciba-Geigy site in McIntosh, Alabama. The company disposed of DDT and other insecticides in the Tombigbee River from 1952 until 1965 (U.S. Environmental Protection Agency, 1992b). This occurred during a period when these disposal activities were legal. In 1983 the EPA put the site on the NPL, and in 1992, Ciba-Geigy signed a $120 million agreement for environmental cleanup of the site. Even though retroactive liability produces large sums of money to pay for cleanup, many citizens, shareholders, and corporations question its fairness.

The Superfund program allocates the costs of cleanup to the private sector to the maximum extent possible by means of the polluter pays principle (see Chapter 6). There is broad public support for this because people think it is fair. To a degree, the Superfund program has been shielded from competing against other domestic programs for general revenue in an era of cost cutting and budget deficits by appealing to this principle. Yet there is often an extended period between the release of toxic substances to the environment and the actual cleanup. Often the parties that caused the contamination or benefited from it are not the same as those being asked to bear the cleanup costs. With retroactive liability, firms that obey all laws and take great care in hazardous waste disposal may be liable for cleanup (see Chapter 5). Joint and several liability can cause firms to be liable for contamination at sites where they sent waste, but where other PRPs caused the contamination through poor disposal practices.

The fairness issue hinges on the questions of who benefited in the past and who pays in the present. Shareholders, managers, and workers of firms who improperly disposed of toxic or radioactive substances in the past benefited from not paying the costs of proper disposal. In addition, past consumers enjoyed lower product prices because some producers did not pay these costs. The Superfund program forces current shareholders, managers, workers, and consumers to pay for the environmental cleanup.

Perhaps the least fair situation occurs at Superfund sites where there are "orphan PRPs," which refers to potentially responsible parties whom the government can no longer find or who are insolvent. Sometimes these orphan PRPs caused most or all of the contamination at a site, yet the other PRPs must pay for the cleanup. PRPs with "deep pockets" are especially at risk in Superfund cleanups associated with orphan PRPs (see Chapter 6).

There are also contaminated sites that are called "orphan sites," for which no responsible party can be identified. An example is the Lansdowne Radioactive Residence site in Pennsylvania, where a physics professor operated a "mom and pop" radium reprocessing laboratory in the basement of his home from 1924 until 1944 (U.S. Environmental Protection Agency, 1992c). EPA added the site to the NPL in 1985, began remediation in 1988, completed cleanup in 1989, and delisted the site in 1991. Cleanup crews hauled away 1338 metal shipping boxes containing 5539 tons of contaminated rubble and soil to a secure radium disposal site in Utah. EPA used $11.6 million dollars from the trust fund to pay for the cleanup because it was unable to identify any responsible parties.

The taxes that finance the trust fund, which is the heart of the Superfund program, are unfair in terms of the polluter pays principle, for they do not distinguish between the guilty and the innocent. Taxes on petroleum and chemical feedstocks apply to all producers whether or not they have ever caused any contamination. To be fair, such taxes should be based directly on past, present, or future hazardous waste pollution episodes.

There are alternative taxes that some analysts claim would be more fair than the current system. The most often discussed alternative is a tax on hazardous waste products, which would encourage firms to minimize wastes in order to minimize taxes, thus helping to prevent future contamination and cleanup problems. Yet two primary problems are that they could encourage midnight dumping and produce a smaller and less certain revenue stream than the present taxes.

In evaluating the fairness of the polluter pays principle, it is also important to consider its indirect effects. Superfund's potent legal powers to enforce the polluter pays principle influence the future behavior of firms and individuals, primarily because the great expense and criminal penalties associated with Superfund enforcement are powerful deterrents to activities that might produce contaminated sites. Efforts to revise a program like Superfund must be careful to ensure that the first priority is for policies that effectively prevent new contaminated sites.

Some critics claim that the way the program operates is not fair to people who own property near Superfund sites. Property values decline or fail to increase as fast for properties near groundwater contamination and Superfund sites compared to properties distant from contaminated sites (see Chapter 6). The decline in value is greatest when EPA adds a site to the NPL, which generates substantial publicity (Hird, 1994, p. 129).

In theory, if real estate markets functioned properly or efficiently, the decline in property value should disappear following cleanup of the site. The Krysowaty site in Hillsborough Township, New Jersey, suggests that in some places real estate markets do function properly (U.S. Environmental Protection Agency, 1992d). EPA placed this site on the NPL in 1982 because of contamination from five years of uncontrolled dumping. The township lowered the property taxes of 68 homes near the site, acknowledging the decreased values resulting from publicity about the Superfund site. The remediation team completed cleanup activities in 1986 and EPA delisted the site in 1989. A local real estate agent claims that "the fear of contamination has subsided and neighboring properties have returned to full market value" (U.S. Environmental Protection Agency, 1992d).

Because environmental cleanup is complete for only a small number of Superfund sites, research results are not clear concerning what really happens to the values of property near Superfund sites after cleanup is complete. Some research suggests that real estate markets do not function well when toxic contamination is present (Page and Rabinowitz, 1993).

Yet there is also the possibility of windfall profits on the value of land near Superfund sites. In this case, windfall profits might accrue to a property owner who purchased land at a low value because of the contamination and then benefits from a rise in the value of the property. Some people consider this a windfall profit because it accrues without any action on the part of the owner. In this situation, the Superfund program causes an environmental cleanup that may result in higher land values. At issue here is the fairness of such windfall profits. Some believe these profits should accrue to the government, which conducted the cleanup (see Chapter 11). Others think that the entrepreneurial activity of purchasing contaminated land or land near a contaminated site is a real risk that warrants any subsequent profits.

Other analysts criticize the fairness of the Superfund program in terms of the spatial distribution of its costs and benefits. One study investigated where the program money was collected and where it was ultimately spent. It found disproportionate penalties for some regions of the country and disproportionate benefits for other regions (McNiel *et al.*, 1988).

In making environmental policy, we must consider not only the magnitude of any risks but also who is bearing the risk. Some argue that rational risk management results in inequity: the poorest and racial minorities are particularly exposed to the risks (Hird, 1994, p. 117). Contaminated sites are highly concentrated in certain locations, especially in urban industrial areas and in remote mining locations. One study found a positive correlation between Superfund sites and counties with a high percentage of nonwhites, although the study had serious methodological problems because of the size of the units of analysis (Hird, 1994, p. 134). The issue of whether the disadvantaged populations near the Superfund sites would benefit or suffer from Superfund activities was not clear. There are countless unfair situations where the poor and racial minorities are most likely to be exposed to contamination risk, and there is a strong moral argument that environmental policy has an obligation to remediate sites for those who cannot afford to escape the risks (World Bank, 1992, p. 83).

CONCLUSIONS

The U.S. Superfund program has changed since its inception in 1980 and will continue to evolve. Yet the overall direction of the changes is not clear. One side advocating change to the Superfund program includes the concerned public and environmental groups who want the program to identify and remediate more contaminated sites. On the other side of the debate are business interests who believe that the United States must scale back the Superfund program or it will bankrupt the country and certainly undermine our competitive position in the world economy. Somewhere in the middle is the federal government which has conducted numerous studies of how to improve the Superfund program. Table 8.4 presents a list of proposed changes, many of which date from 1988 and are still under discussion today. Some change is certain because everyone believes that the Superfund program can operate better.

Financing of the Superfund program is the critical issue. Almost everyone agrees that the government can reform the program to more efficiently use trust fund resources. Some believe that the present financing mechanisms can be modified to function better and more fairly. Others believe that radical reform is necessary, particularly the rejection of retroactive, strict, and joint and several liability as the basis of funding Superfund cleanups.

Despite the criticism and controversy over Superfund, it is making progress in addressing a significant environmental problem and it has wide public support. In terms of its accomplishments, it may be the most success-

TABLE 8.4

PROPOSED STRATEGIC INITIATIVES TO IMPROVE THE U.S. SUPERFUND PROGRAM[a]

Setting Cleanup Priorities and Goals
1. Set priorities on basis of current or future risks
2. Establish a federal site discovery program
3. Use environmental criteria to eliminate sites at PA and SI screening stages
4. Remove range of acceptable risk objective
5. Establish national minimum cleanup standards
6. Define and limit meaning of permanent cleanup

Developing Workers and Technologies
7. Reduce dependency on contractors and expand EPA workforce
8. Establish a heirarchy of cleanup technologies and methods
9. Restrict use of groundwater cleanup technology
10. Establish generic site assistance program, including expert systems
11. Establish technologies assistance program
12. Better define mission of SITE technology demonstration program

Improving Government Management
13. Use generic site classification
14. Limit responsible parties to implementation of remedies
15. Reexamine financing and enforcement of liabilities to improve environmental performance
16. Strengthen EPA headquarters direction and oversight of regional implementation
17. Commit to a permanent Superfund program
18. Establish an all-inclusive list of cleanup sites in the United States
19. Begin examination of moving Superfund implementation outside of EPA

[a]Modified from U.S. Congress (1988, Table 1.1).

ful environmental cleanup program in the world. Given the extent of the contamination and cleanup problem in the United States, the Superfund program is likely to continue operation for many years, even as it undergoes many modifications as the government attempts to improve its operation.

References

Acton, J. P., and L. Dixson. 1992. *Superfund and Transaction Costs: The Experiences of Insurers and Very Large Industrial Firms.* Santa Monica, CA: Rand.

Crandall, R. 1988. What ever happened to deregulation? In D. Boaz (ed.), *Assessing the Reagan Years.* Washington, DC: Cato Institute, 284 pp.

Department of Natural Resources Protection. 1994. *Bi-annual Inventory Report of Contaminated Locations in Broward County, Florida.* Fort Lauderdale, FL: Broward County Department of Natural Resources Protection, 97 pp.

Environmental Damage Valuation and Cost Benefit News. 1994. Vol. **1**(1), 1.

Herman, S. 1995. Superfund administrative reforms will result in faster cleanup, more effective program. *Nat. Environ. Enforcement J.* **10**(2).

Hird, J. A. 1994. *The Political Economy of Environmental Risk: Superfund*. Baltimore, MO: Johns Hopkins University Press, 316 pp.

McNiel, D., A. Foshee, and C. Burbee. 1988. Superfund taxes and expenditure: Regional redistributions. *Rev. Reg. Stud,* Winter, 4–9.

National Research Council. 1991. *Environmental Epidemiology: Public Health and Hazardous Waste Sites*. Washington, DC: National Academy Press.

Page, G. W., and H. Rabinowitz. 1993. Groundwater contamination: Its effects on property values and cities. *J. Am. Plann. Assoc.* **59**(4), 471–479.

Portney, P. 1988. Reforming environmental regulation: Three modest proposals. *Issues Sci. Technol.* Winter, 74–81.

Probst, K., and P. Portney. 1992. *Assigning Liability for Cleanups: An Analysis of Policy Options*. Washington, DC: Resources for the Future.

Romano, A., and J. Simon. 1993. Superfund accelerated cleanup model. *Remediation* **3**(2), 247–257.

Russel, M. E., W. Colglazier, and M. English. 1991. *Hazardous Waste Remediation: The Task Ahead*. Knoxville, TN: Waste Management Research and Education Institute.

U.S. Congress, Office of Technology Assessment. 1988. *Are We Cleaning Up? 10 Superfund Case Studies* (OTA-ITE-362). Washington, DC: U.S. Government Printing Office.

U.S. Congress, Office of Technology Assessment. 1989. *Coming Clean: Superfund's Problems Can Be Solved* . . . (OTA-ITE-433). Washington, DC: U.S. Congress.

U.S. Environmental Protection Agency. 1987. *A Comparative Assessment of Environmental Problems: Overview Report,* Vol. 1. (Office of Policy, Planning and Evaluation.

U.S. Environmental Protection Agency. 1990a. *The Revised Hazard Ranking System: Background Information*. Office of Solid Waste and Emergency Response Publ. 9320.7-03FS. Washington, DC.

U.S. Environmental Protection Agency. 1990b. *Field Test of the Proposed Revised Hazard Ranking System (HRS)* (EPA 540/p-90–001). Washington, DC.

U.S. Environmental Protection Agency. 1990c. *An Analysis of State Superfund Programs: 50-State Study, 1990 Update* (EPA 540/8-91-002). Washington, DC.

U.S. Environmental Protection Agency, Science Advisory Board. 1990d. *Reducing Risk: Setting Priorities and Strategies for Environmental Protection* (SAB-EC-90-021). Washington, DC.

U.S. Environmental Protection Agency. 1992a. *Superfund at Work. Valley of the Drums Cleanup: A Superfund Benchmark* (EPA 520/F-92-006). Solid Waste and Emergency Response, Washington, DC.

U.S. Environmental Protection Agency. 1992b. *Superfund at Work, EPA Wins Ciba-Geigy's Full Cooperation to Clean Up Alabama Site* (EPA 520/F-92-017). Solid Waste and Emergency Response, Washington, DC.

U.S. Environmental Protection Agency. 1992c. *Superfund at Work. EPA Completes*

Cleanup of Nation's Only Residential Superfund Site (EPA 520/F-92-011). Solid Waste and Emergency Response, Washington, DC.

U.S. Environmental Protection Agency. 1992d. *Superfund at Work. Krysowaty Farm Cleaned Up* (EPA 520/F-92-016). Solid Waste and Emergency Response, Washington, DC.

U.S. Environmental Protection Agency. 1992e. *Superfund at Work. Cleanup Begins at Well G&H, One Year after Landmark New England Settlement* (EPA 520/F-92-015). Washington, DC.

U.S. Environmental Protection Agency. 1993a. *Superfund at Work. The Superfund Removal Team* (EPA 520/F-93-014). Solid Waste and Emergency Response, Washington, DC.

U.S. Environmental Protection Agency. 1993b. *Superfund at Work. EPA and Olin Clean Up Triana Site: A Major Victory for the Environment* (EPA 520/F-93-001). Washington, DC.

U.S. Environmental Protection Agency. 1993c. *Superfund at Work. Accelerated Cleanup at Tri-State Plating* (EPA 520/F-93-009). Washington, DC.

U.S. Environmental Protection Agency. 1993d. *Superfund at Work. Environmental Victory at Army Creek Landfill* (EPA 520/F-93-015). Washington, DC.

U.S. Environmental Protection Agency. 1993e. *Superfund at Work. Innovative Technology Used to Clean Up French Limited* (EPA 520/F-93-004). Washington, DC.

U.S. Environmental Protection Agency. 1993f. *Superfund at Work. Innovative Technology Used to Restore Environment* (EPA 520/F-94-001). Washington, DC.

U.S. Environmental Protection Agency. 1993g. *Superfund at Work. EPA Gains Cooperation of 86 Parties to Clean Up Coal Creek* (EPA 520/F-93-011). Washington, DC.

U.S. Environmental Protection Agency. 1995. *Land Use in the CERCLA Remedy Selection Process,* OSWER Directive No. 9355.7-04. Washington, DC.

U.S. General Accounting Office. 1991. *Hazardous Waste: Limited Progress in Closing and Cleaning Up Contaminated Facilities* (GAO/RCED-91-79). Report to the Chairman, Environment, Energy, and Natural Resources Subcommittee, Committee on Government Operation, House of Representatives. Washington, DC: U.S. Government Printing Office.

U.S. General Accounting Office. 1992. *Superfund: Problems with the Completeness and Consistency of Site Cleanup Plans* (GAO/RCED-92-138). Washington, DC: U.S. Government Printing Office.

U.S. General Accounting Office. 1993a. *Superfund: Progress, Problems, and Reauthorization Issues* (GAO/T-RCED-93-27). Washington, DC: U.S. Government Printing Office.

U.S. General Accounting Office. 1993b. *Superfund: Cleanups Nearing Completion Indicate Future Challenges* (GAO/RCED-93-188). Washington, DC: U.S. Government Printing Office.

U.S. General Accounting Office. 1995a. *Superfund: System Enhancements Could Improve the Efficiency of Cost Recovery* (GAO/AIMD-95-177). Washington, DC: U.S. Government Printing Office.

U.S. General Accounting Office. 1995b. *Superfund Program Management* (GAO/HR-95-12). Washington, DC: U.S. Government Printing Office.

U.S. General Accounting Office. 1996. *Superfund: Implications of Key Reauthorization Issues* (GAO/RCED-96-145). Washington, DC: U.S. Government Printing Office.

U.S. Senate, Committee on Environment and Public Works. 1983. *A Legislative History of the Comprehensive Environmental Response, Compensation, and Liability Act of 1980,* Vol. 1, Serial No. 97-14. Washington, DC: U.S. Government Printing Office.

World Bank. 1992. *World Development Report 1992: Development and the Environment.* New York: Oxford University Press.

CHAPTER 9

Contamination and Environmental Cleanup in Western Europe

INTRODUCTION

The countries of Western Europe are all taking steps to contend with their contamination problems, which vary substantially between countries. There is even greater variation, however, in how they respond. The differences in contamination are a function of the level of industrial activity and especially the nature of government policies to control pollution. All European countries have had pollution control policies in place for many years, but implementation of these policies has not been consistent. For instance, Italy has good environmental legislation in many respects, but only the government may initiate legal proceedings and it has hardly ever done so (Bianchi, 1994). The countries of Central and Eastern Europe, which had communist governments in the decades following World War II, did not enforce pollution control laws and established economic incentives that encouraged pollution. Accordingly, the extent of the contamination and environmental cleanup problem is substantially different in Western Europe than in the former communist countries.

THE EUROPEAN UNION

The dominant political force in Western Europe is the European Union (EU). The Maastricht Treaty of 1991, which became effective in 1993, created the European Union from what had been the European Community. The European Union is now attempting to establish a common policy concerning environmental cleanup for its member states and for all businesses operating therein. European Union policies strongly influence the small number of Western European countries that are not members. The countries of the European Free Trade Area—Iceland, Norway, Liechtenstein, and Switzerland—have negotiated an agreement with the European Union that covers some environmental topics.

Policy toward environmental cleanup and the reuse of contaminated brownfield sites is evolving in the countries of the European Union, and at present every member-state has their own national environmental policies. Yet there is considerable variation in the legislation concerning environmental cleanup and even greater variation in the extent to which governments implement these policies. The European Union is moving toward enacting legislation that would supersede national policies and therefore apply to all member-states.

The European Union has grown in both number of member-states and political responsibility. It had its start in the European Economic Community, which six countries established with the Treaty of Rome in 1957. From this original purely trade organization, the European Union has grown to a level of government that approaches a federal government of the independent European member-states. The European Union has its own European Parliament, a Council of Ministers, and the European Commission. This trend toward a federal government for Europe is likely to continue as the European Union moves toward a single currency within the next decade. In January 1995, it grew to its present size of 15 countries: Austria, Belgium, Denmark, Finland, France, Germany, Greece, Ireland, Italy, Luxembourg, The Netherlands, Portugal, Spain, Sweden, and the United Kingdom.

The European Union has governmental powers that make it more like an independent country and different from most international organizations. Most significantly, laws established by the European Union take precedence over national laws enacted by the member-states. For example, if the European Union establishes a law on environmental cleanup of soil and groundwater contaminated with toxic substances, then that law would apply in all the member-states. This is different from other international organizations such as the North American Free Trade Association (NAFTA), which depends on its member-states to enact laws that further NAFTA ends.

The European Union would like to establish a single policy on environmental cleanup that applies to all the member-states, the main motivation being to protect economic competitiveness. Competition for the location of economic activity increased when the European Union established the Single European Market at the end of 1992. This economic competition was between cities, regions, and countries in all of the member-states. Environmental policies may affect the locational decisions of firms seeking to open new facilities or to relocate.

Environmental policies vary significantly between the member-states. Some countries fear that policies that force expensive environmental cleanups will cause industries to locate in other EU countries that have weaker policies. The Netherlands, Germany, and some other countries have strong policies to force cleanup of contaminated sites, which can make these countries expensive places to operate businesses. Some member-states, especially those countries along the southern tier of the EU, have policies that are less effective and less costly. The European Union would like to have a common policy that would not distort economic decisions and that would create a "level playing field" for economic location decisions.

A common environmental cleanup policy for the European Union would also have other benefits, such as long-term economic benefits by eliminating contamination that might become an even greater burden in the future and thus improving the quality of life. In addition to reducing human health risks and improving ecosystem health, these policies would help in achieving the redevelopment of brownfield sites, which is an important issue in these densely populated countries that want to avoid urban sprawl and the loss of open space. A common policy would also help the European Union in its dealings with the countries of Central Europe who would like to join the Union, for this would force these countries to establish financing mechanisms to pay for cleanups and begin cleaning up their own contaminated sites.

EUROPEAN UNION ROLE IN SETTING ENVIRONMENTAL POLICY

The European Union has the legal authority to establish an environmental policy for cleanup and the reuse of contaminated brownfield sites within its member-states. The original European Economic Community was concerned with trade and not environmental issues. The Single European Act of 1987 added the Environment Title to the Treaty of Rome and thereby added environmental concerns to the original focus on trade.

In 1989, the Commission of the European Economic Community issued a *Directive on Civil Liability for Damage Caused by Waste* (COM/89/282), which was its first policy statement on the issue of environmental cleanup.

Much of this directive appears to be influenced by the U.S. Comprehensive Environmental Response, Compensation, and Liability Act (CERCLA) enacted in 1980. The European Union amended this directive in 1991 (COM/91/219), the same year that the member-states ratified the Maastricht Treaty. These directives embody the polluter pays principle that is stated in the Environment Title of the Maastricht Treaty (Title XVI). The directive proposed strict, joint and several liability as the means to raise the funds necessary to pay for environmental cleanup. There are some minor differences with the U.S. Superfund liability, such as in limiting liability to 30 years for old contamination.

In Western Europe in 1993, there was additional activity advocating environmental cleanup policy. The European Commission released its *Green Paper on Remedying Environmental Damage.* Green Papers are issued to promote discussion among the member-states prior to enacting legislation. This Green Paper repeated the arguments of the *Directive on Civil Liability for Damage Caused by Waste* that the Union adopt a common system of civil liability, which would help balance economic competition across countries. Also in 1993, the Council of Europe, a 23-member Western European organization, issued its *Convention on Civil Liability for Damage Resulting from Activities Dangerous to the Environment.* The Council also advocated a common environmental policy for cleanup.

Other factors were also driving environmental awareness, in particular the rise of Green political activism and episodes of toxic substances contamination. Two of the most publicized incidents were the 1976 Seveso, Italy, leak of trichlorophenol contaminated with dioxins and the 1986 chemical fire in Basel, Switzerland, which contaminated the Rhine River. Partly in response to these events, the European Community agreed in 1990 to create the European Environment Agency, which the European Union finally established in Copenhagen in 1994. Presumably in the future, the European Environment Agency will play an important role in developing environmental policy for the European Union. However, the European Commission is responsible for officially proposing all legislation for the European Union, and the Maastricht Treaty allows member-states to adopt more stringent environmental standards than those of EU legislation.

Yet as the European Union has become increasingly active in environmental policy, there is a concern by the member-states to retain some control over policy in their own countries. The term "subsidiarity" embodies the official doctrine of the European Union that member-states should make decisions at the most local level of government that is appropriate for the issue at hand. Subsidiarity implies that each of the member-states should establish all environmental policy that is not Europe-wide in scope.

The issue of subsidiarity is analogous to that of "states rights" in the United States.

Before the European Union can enact legislation, those advocating EU environmental policy must convince member-states that the issue at hand is Europe-wide in scope. In the case of environmental cleanup, most observers assume that an insufficient number of member-states would vote for a common cleanup policy. It is probably only a matter of time before the European Union establishes such a policy, since the expansion to 15 member-states in 1995 limits the ability of a small number of member-states to block legislation.

ENVIRONMENTAL CLEANUP IN THE EUROPEAN UNION

Although there is no common EU policy toward environmental cleanup, there are commonalities in the approach to cleanup and reuse of contaminated sites among the member-states. These commonalities are especially clear when compared to the U.S. approach (see Chapter 8). The governments of member-states accept a substantially greater responsibility to pay for cleanup as compared to the United States, which attempts to force private parties to pay for more of the cleanup. European countries also play a much greater role and are more successful in the reuse of remediated land than is the United States.

Yet the European Union and its member-states follow the polluter pays principle. When a member-state government identifies an individual or firm that has caused toxic substances contamination of soil or groundwater, they attempt to make this firm pay for the cleanup. This is the same approach taken in the United States.

The difference occurs when the government cannot find a polluter capable of paying. In these instances, the European countries generally use government funds to pay for the cleanup, whereas the United States uses extreme legal powers to force private parties (potentially responsible parties) to pay for cleanup. In legal terms, EU countries depend on "fault-based" liability, which requires that an individual or firm be at fault in causing the contamination before the government can force them to contribute cleanup costs.

As discussed in the preceding chapter, the U.S. government uses strict, retroactive, and joint and several liability to cast a net of legal obligation over a wide array of private firms. In the United States, it is possible that the liability system can force minor polluters or even innocent parties to pay for environmental cleanup, particularly when the government cannot identify the polluter or if the polluter is incapable of paying.

ENVIRONMENTAL CLEANUP IN THE FORMER EAST GERMANY

The European Union countries have had a unique opportunity to observe how one of their members, Germany, deals with the environmental cleanup of the former East Germany following reunification in October 1990. The reunited Germany inherited a staggering number of sites in need of cleanup. For example, chemical and other industries from the former German Democratic Republic indiscriminately dumped toxic wastes in abandoned open-cast lignite mines, leaving more than 70,000 hectares of contaminated abandoned mines.

Germany attempts to use the polluter pays principle to obtain funds for environmental cleanup. The now nonexistent communist government of the former East Germany was the owner of all enterprises that created the contamination, so in this case the polluter has literally evaporated into the thin German air.

Following reunification, the German government wanted to assist and speed the integration by cleaning up environmental contamination and stimulating the economy of the former East Germany. The government plans to privatize East German industrial facilities as a way to promote investment. To achieve this, the German government waived most of the liability for the cleanup of contaminated sites and covered the costs from its own funds.

An example of such a cleanup is the electrical transformer plant in Dresden in the former East Germany. Siemens AG owned the plant before World War II, and communists ran the plant for 40 years. There is ample evidence of contamination, but no records of how much or where the industries used or disposed of toxic substances. Siemens bought the plant after reunification and paid an initial, nominal environmental cleanup fee equivalent to $3 million U.S. for a cleanup that is certain to cost many millions of dollars (Litchfield, 1992). The German government will pay for the rest of the cleanup.

REDEVELOPMENT OF CONTAMINATED BROWNFIELD SITES

Another major difference between the European countries and the United States is their approach to the reuse of contaminated sites following clean-up. Western European countries return many more remediated sites to productive use, for many possible reasons. The overriding reason may be that EU member-states are more densely populated than the United States and so have fewer greenfield sites available as a ready alternative to redeveloping brownfield sites. This scarcity of alternative sites is related to the high priority on maintaining greenbelts and open space around European cities. In the United States, with its relatively large supply of readily available greenfield sites, few developers are interested in remediated sites, even when they have excellent infrastructure.

Another important factor is that U.S. governments lack effective control of land use, which allows firms to develop new facilities on greenfield sites instead of redeveloping formerly contaminated brownfield sites. Even when local U.S. governments form public–private partnerships to promote the reuse of remediated sites, they are limited in their capabilities compared to European governments. The latter have greater control of land use and thus a greater capability to acquire and assemble land and to broker real estate deals to assist the redevelopment process. In the United States, governments have less ability to control land use and must often compensate private landowners for any restrictions they impose.

The European Union also has programs that explicitly promote place-based economic development. This economic assistance is a subsidy to promote economic growth in selected places, often in urban areas. Place-based subsidies are designed to prevent the growth of urban blight and to promote the reuse of brownfield sites. In 1994, the European Union spent 23 billion ECUs, approximately $25.2 billion U.S., on such "regional aid" (*The Economist,* 1994). The United States spends much less on providing place-based assistance to urban areas, apart from limited block grant funds that the federal government distributes to all municipalities.

Liability systems exert a strong influence on the reuse of land after environmental cleanup. The fault-based system of liability used in EU countries allows a clear determination of the liability for a contaminated site that facilitates redevelopment. For example, the owner or a local government may accept all of the liability for a site, thereby enabling a developer to purchase the site for reuse free of liability concerns.

In contrast, the U.S. application of liability greatly limits the possibility of indemnification and therefore the incentive for developers to risk the purchase and redevelopment of a remediated site. Liability under CERCLA makes any owner potentially liable for environmental cleanup, thus discouraging potential developers. CERCLA liability also makes future purchasers of remediated sites potentially liable for future environmental cleanups if additional contamination is discovered or if the government tightens cleanup standards. In Europe, a developer who purchases a remediated site is not liable for any contamination that they do not cause.

Unlike European governments, which are more involved and more successful in redeveloping remediated sites, the U.S. government stays relatively uninvolved in such efforts and depends on the private market to return the land to productive use. This especially applies to the most severely contaminated sites in the Superfund program; it is less accurate regarding contaminated sites in the various mini-Superfund programs operated by the individual states.

CONCLUSIONS

The countries of Western Europe are among the most successful in controlling toxic contamination and conducting environmental cleanups of contaminated sites. This is especially true of the countries of the European Union.

The European Union is currently developing an environmental cleanup policy and is making significant progress toward greater cooperation in establishing common policies to deal with common problems among its member-states. It is already exerting a tremendous influence on the countries of Central and Eastern Europe. Because these countries aspire to someday join the European Union, they feel pressure to establish similar policies and standards. This can have a tremendously beneficial effect on their own toxic contamination and cleanup problems. Even though the European Union countries are among the world's leaders in addressing contamination and environmental cleanup, all of these countries are still experimenting and searching for better policies.

References

Bianchi, A. 1994. The harmonization of laws on liability for environmental damage in Europe: An Italian perspective. *J. Environ. Law* **6**(1), 21–42.

The Economist. 1994. March 26, p. 66.

Litchfield, R. 1992. The $20-billion question: Who pays. *Can. Bus.* September, p. 15.

CHAPTER 10

THE NETHERLANDS CASE STUDY

INTRODUCTION

Because of its high population density, scarcity of undeveloped land, high concentration of manufacturing, and high utilization of groundwater for drinking, The Netherlands has become one of the leading countries in developing environmental cleanup policy. Estimates for the number of suspected contaminated sites in The Netherlands range from 200,000 to 600,000, approximately 99% of these sites being of industrial origin (Soczó *et al.*, 1992; Business Roundtable, 1993).

The Netherlands was the first European country to enact policy specifically directed at addressing the contaminated sites and environmental cleanup problem, with its earliest policies dating from 1980 (Visser, 1993). Their policy approach has influenced several other European countries. They also have an innovative voluntary cleanup program for contaminated sites that operates as a public–private partnership.

The Netherlands, along with other continental European countries, has a legal system that is descended from Roman law. The Roman law-based legal systems are different from those of the United Kingdom, the United States, and some other countries, whose legal systems are descended from common law.

In Roman law countries, municipal plans for the use of land are legally binding. In common law legal systems, land use plans are discretionary and intended to guide land use, but are not legally binding. The importance of this distinction is that in Roman law countries, municipalities have a greater ability to determine which properties will receive a cleanup and especially which properties will be redeveloped after remediation.

A distinctive feature of land markets in The Netherlands makes cleanup and redevelopment of contaminated sites different from that of many other countries. Dutch municipalities play the unique role in urban land markets of making land available to developers. When someone proposes a development or redevelopment, the municipality acquires the land, services the land to make it suitable for the proposed development, and then offers it to the developer (Needham *et al.*, 1993, p. 23). Servicing the land usually means providing appropriate public infrastructure such as roads, water, and sewers. However, when municipalities service contaminated land, servicing also entails working with the Dutch Ministry of Housing, Physical Planning, and the Environment to provide the environmental cleanup. Local governments often retain ownership of the land and provide only a lease to the developers. Thus local government in The Netherlands plays a key role in environmental cleanup that is not present in most other countries.

PROCESS FOR RECLAIMING CONTAMINATED LAND

In The Netherlands, the first step in the cleanup process occurs when municipalities identify contaminated sites and the province lists the sites. Each year the province presents to the Dutch Ministry of Housing, Physical Planning, and the Environment a five-year cleanup program for contaminated sites. In consultation with the municipalities, the province develops its own priorities for site cleanups and is responsible for implementing the cleanup plan. For government-funded cleanups, the Dutch Ministry provides about 90% of the cleanup funding, with about 10% coming from the municipality (Kingsbury and Bingham, 1992, p. 215). The government will then bring legal action to recover the cost from the responsible party.

Following identification and listing of a contaminated site, there is a three-stage process: (1) soil investigation, (2) exposure investigation, and (3) cleanup investigation (Visser, 1993, p. 52).

The first stage involves efforts to determine if there is sufficient contamination to warrant remedial action. This investigation includes analysis of past uses of the site and searches for potential sources of contamination. It conducts sampling and analyzes soil and groundwater. The government will not require remedial action if concentrations of chemicals do not exceed the Dutch List A-values.

This value represents a concentration of the substance that is considered "natural." The Netherlands uses the Dutch List values—levels A, B, and C—to determine which sites require remediation and the level of remediation needed. These levels are described in greater detail in a following section.

If concentrations of any chemicals from the samples exceed the Dutch List B-values, then investigators must complete additional sampling and analysis (Visser, 1993, p. 52). The temporary replacement of the B-value under the 1994 legislation is the value (A + C) / 2, where A and C represent the respective A-list and C-list values for the substance under investigation.

The government requires the second stage of the process for restoring contaminated sites when chemical concentrations exceed the Dutch List C-values. Such concentrations indicate the necessity of environmental cleanup, at which time the site is added to the cleanup list. Exposure investigation evaluates actual and potential exposure of human populations to toxic chemical contamination on the site and the ecological effects. It also determines the urgency of remedial action and the cleanup timetable.

The third stage of the process—cleanup investigation—establishes the remedial action plan. Investigators use information from site-specific investigations and feasibility analyses to determine this plan, and consider present or intended future land use. In The Netherlands, the goal of remediation is multifunctionality, which is restoration of a site so it can support all potential land uses. However, in some cases, achieving multifunctionality is not feasible and containment of contaminants on-site may be the goal of the remedial action plan (see the following discussion of the 1994 legislation).

LEGISLATIVE HISTORY IN THE NETHERLANDS

The Netherlands has enacted a series of legislation to address toxic contamination and environmental cleanup, and their policy continues to evolve. More than most countries, The Netherlands has an evolutionary style of policy development. Industries and individuals often implement new policy directions by voluntary means before government enacts new legislation. The following discussion attempts to cover the main policy directions rather than be a complete timeline of legislation.

Interim Soil Cleanup Act

In 1981, The Netherlands adopted the Interimwet Bodemsanering (Interim Soil Cleanup Act). The government intended this "emergency law" to fund the cleanup of severely contaminated land for the next 10 years. Wide publicity of contamination in the town of Lekkerkerk contributed to its passage (see the Ledderkerk case study later in this chapter). This act anticipated as many as 350 cleanups at an estimated total cost of one billion Dutch guilders (NLG) (approximately $578 million U.S.).

The Interim Soil Cleanup Act had the following components: the province implements the act by carrying out the cleanup and paying for the cleanup, and the central government initiates legal proceedings against the polluter based on environmental tort to recover the cost of cleanup. Thus the act is based on the polluter pays principle.

The goal of environmental cleanup is multifunctionality, which includes being able to build diverse human structures on it, to extract groundwater, to produce crops, and to provide raw materials. According to the multifunctionality standard, the present use of the land does not determine the degree of cleanup. Instead, the land should be sufficiently clean for any land use in the future. In practice, remedial action plans often neither aim for nor achieve multifunctionality. Only 8% of cleanup actions under the Interim Soil Cleanup Act achieved the goal of multifunctionality (van Arnheim, 1994).

Soil Cleanup Guidance Note

In 1983, The Netherlands government issued Leidraad Bodembescherming (Soil Cleanup Guidance Note) concerning contaminated land (Ministry of Housing, 1983). This Guidance Note by itself had no judicial status; however, local government adopted it, thereby making it official policy. The Guidance Note establishes three levels of contaminated land: harmful contamination, urgent contamination, and postcleanup multifunctionality.

Harmful contamination represents severe danger for public health or the environment. The Guidance Note specifies three levels of harmful contamination, with each level determined by the concentration of hazardous substances present. Other governments, for instance, West Germany and Poland, adopted these lists, which are known as the Dutch List. The three levels are:

- A-level—the reference level that requires no action;
- B-level—contamination requires further environmental assessment of the site; and
- C-level—the intervention level at or above which cleanup is required.

Each hazardous substance listed in the Guidance Note has its own A-, B-, and C-level concentrations for the substance in soil and different levels for the substance in groundwater. The Dutch list contains values for metals, inorganic constituents, aromatic compounds, polycyclic hydrocarbons, chlorinated hydrocarbons, toxic pesticide products, and other contaminants.

The government establishes the Dutch List values using different criteria. The A-values for soil correspond to the "mean" background concentration found in The Netherlands or detection levels for substances that are not naturally occurring. For groundwater, the government set the A-values at concentrations used to protect drinking water in The Netherlands and the European Community. The government set the B-level concentration at the midpoint of the A- and C-level concentrations for each substance. The C-level concentrations are based on the toxicity and persistence of the contaminant.

The A-, B-, and C-level values are intended to be used in conjunction with site-specific information on land use and hydrogeologic conditions. However, these values are quantitative values and the land use and hydrogeologic information is qualitative. The result is that officials base their decisions about the need for cleanup mostly on the A, B, and C-level values (Visser, 1993, p. 48). It is important to note that the Dutch government uses these values for site assessment, but does not intend them to be standards for site cleanup. In practice, this means that government will remediate contamination to below A-level concentrations, the no-action level (van Arnheim, 1994).

Urgent contamination refers to land with harmful contamination that the provincial government designates as "urgent", usually if it threatens public health. Serious groundwater contamination often triggers this designation. When government makes such a determination, legislation requires environmental cleanup within specified periods: 5 years under the Interim Soil Cleanup Act of 1981 and 4 years under the Soil Protection Act of 1994 (see the following). Experience in The Netherlands and in other countries suggests that these time frames are not realistic.

Soil Protection Act

The adoption of the Wet Bodembescherming 1986 (Soil Protection Act) represented the first legislation to effectively address the contamination and cleanup problem in The Netherlands. Under the Soil Protection Act, the Dutch Ministry of Housing, Physical Planning, and Environment was responsible for establishing policy to protect land from contamination. In 1994, The Netherlands included the Interim Act as a chapter in the Soil Protection Act. The Ministry appointed a Technical Soil Protection Committee to review proposals and make recommendations.

The basic policy approach of The Netherlands is that only severe cases of soil pollution will be remediated within the scope of the national soil cleanup regulations (Visser, 1993, p. 48). The provinces and a few large cities with responsibilities comparable to the provinces implement the policy. There is also a government-operated Soil Cleaning Service Center, which is in charge of all environmental cleanups that use any government money (Visser, 1993, p. 45).

EXPERIENCE WITH POLICY AND SPECIFIC ACTIONS

By the late 1980s, experience with contamination and cleanup in The Netherlands had revealed that the problem was larger than envisioned in 1981, when the government enacted the Interim Soil Cleanup Act. That act had anticipated as many as 350 cleanups at an estimated total cost of one billion NLG. By 1990, the number of contaminated sites had risen to at least 70,000, and the cost was estimated in the range of 50 to 100 billion NLG ($29 to $58 billion U.S.). It was clear that the Interim Soil Cleanup Act was not capable of contending with the unexpected magnitude of the contamination problem that existed in The Netherlands.

In response, the government established a tax on firms that treat, store, or dispose of chemical wastes collected from primary sources. The charge was based on the volume of waste that the firms handled. The government intended for this tax to raise substantial funds to pay for disposal, reduction, and prevention of contaminated wastes, yet it failed to generate the targeted amounts of revenue. It also created the unintended incentive for firms to export hazardous waste from The Netherlands or to engage illegal disposal activities (Bernstein, 1993, p. 59). The Netherlands replaced this charge with a fuels tax in 1988.

In the early 1990s, the highest court precipitated a crisis in national environmental policy. The court decided that firms causing contamination before that act was illegal could not be liable for environmental cleanup. Actions prior to 1975 cause no tort basis for legal action to recover the cost of cleanup from the party responsible for the contamination (P. van Arnheim, facsimile, March 8, 1994). The court based its ruling on the lack of explicit government concern for soil pollution prior to January 1975 (see *Staat v. Van Amersfoort 1990* and *Staat v. Azko 1992*). That year is significant because the court ruled that the government cannot apply the polluter pays principle to events prior to 1975. This court decision undermined aspects of the Interim Soil Cleanup Act, which depended on tort legal proceedings to recover the cost of environmental cleanup.

In 1991, private oil companies and the petrol distributors association in The Netherlands established a program to fund environmental cleanups. This program specifically addressed the cleanup of contaminated sites resulting from leaks and spills at service stations (Business Roundtable, 1993, p. 78) and allowed the government to inspect for contamination at the country's 6200 stations. This program is funded by consumers via an increase in the price of petrol.

In 1992, The Netherlands initiated a public–private partnership called Operation Cleaning Up Contaminated Locations in Use, in which private companies voluntarily remediate their contamination. However, the government is active in selecting sites on which there is a risk of contamination and urges companies operating facilities on these sites to conduct an environmental assessment to determine if toxic or hazardous contaminants are present. The companies then negotiate a remedial action plan with the government. If companies do not voluntarily conduct a cleanup, then the national government will report the situation to the province, who will issue a cleanup order.

Also in 1992, The Netherlands acted to control the transfer of ownership of contaminated sites. Such transfers now require government approval and must meet satisfactory guaranties and financial securities for cleanup. If the parties do not supply such guaranties, the government denies the transfer (Parliamentary documents, No. 22727, introduced by the government on August 25, 1992, from Business Roundtable, 1993, p. 75). Under the Housing Act, municipalities can deny a building permit until the developers survey soil and groundwater to determine if contamination is present (Business Roundtable, 1993, p. 77).

Soil Protection Act of 1994

With adoption of the Wet Bodembescherming 1994 (Soil Protection Act), parliament integrated aspects of the Soil Protection Act of 1980 and the Interim Soil Cleanup Act of 1991. Also referred to as the Extended Soil Protection Act, it was originally considered in 1992, but the Senate continued debate of the Act until 1994 when it was passed in amended form and signed by the Queen in the spring. This legislation specifies a three-stage hierarchy for the process of environmental cleanup of contaminated sites:

1. The highest preference is for the polluter to voluntarily clean up the contaminated sites.
2. The second preference is for an enforced survey of the contaminated sites followed by a "cleanup order", which the government delivers to the owner and any user of the land. The cleanup order is a significant innovation and may play an important role in speeding up the cleanup process.
3. The third and least preferred way for cleanups to occur in The Netherlands under this policy is for the government to conduct the cleanup and attempt to recover the cost using environmental tort legal action.

This third approach is the approach advocated in the Interim Soil Cleanup Act of 1981.

The Soil Protection Act of 1994 made several significant changes to earlier legislation. The government expanded A-level and C-level concentrations and revised them on the basis of ecotoxicological models and all potential human exposure routes. It also eliminated the B-levels.

This legislation specifically covers contamination existing before 1975 and bases all questions of historical liability on fault and the individual assessment of cases. Accordingly, the government can take action against both the polluter and the property owner. The provinces as well as the national government are authorized by the Soil Protection Act of 1994 to take legal action to recover the cost of environmental cleanup. The property owner may be liable even if they did not cause the contamination, so long as they fail to prove that they are an "innocent owner" (see the following exceptions). A property owner can be charged for the cost of cleanup, although the cost to a property owner who did not cause the contamination is limited to the difference in the value of the property as clean property and its value as contaminated property.

The province will not issue a cleanup order to an innocent owner, defined as an owner of a contaminated site who was not responsible for or involved in causing the contamination. Under this act, a landowner must prove four conditions to be considered an innocent owner:

1. That the owner played no role in causing the contamination;
2. That the owner had no business involvement with the polluter, such as a lease or some form of loan or investment;
3. That the owner was unaware of the existence of the hazardous substances on the property; and
4. That there was no reasonable expectation that the owner would know of the contamination. Thus the government may send a cleanup order to a property owner who failed to conduct an environmental assessment of the property despite risk of contamination from a known source or previous activity on the site.

In late 1994, the Dutch Senate eliminated the general policy stating that owners who could not prove they were innocent would be liable for cleanup of contamination.

THE DUTCH APPROACH

The Netherlands bases environmental liability on fault, that is, the person responsible for the release of the toxic chemical is liable. This is the polluter pays principle; the responsible party is liable for any costs incurred by national, province, or municipal governments in environmental cleanup of soil or groundwater and for third-party claims.

There is strict liability, but no retroactive liability for pollution that took place before legislation made the polluting activity illegal (prior to 1975). This approach should allow for clear identification of the responsible (liable) parties, eliminate the need for multiple parties insuring for the same risk, and allow insurance companies to assess risk more easily (Business Roundtable, 1993, p. 75). The issue of liability for land to which several parties contributed to the contamination is not yet clear. Each party has some liability, but administrative cleanup orders may have the same effect as joint and several liability. This could cause any one of the responsible parties to be liable for the entire cost of cleanup (Business Roundtable, 1993, p. 78). The Business Roundtable report (1993) provides a detailed description of liability under various legislative acts.

The Soil Protection Act of 1994 emphasizes multifunctionality by specifically prohibiting cleanup that does not achieve this goal. In fact, the act prohibits provinces from approving cleanup plans that do not strive for multifunctionality. However, there are four exemptions from the requirement of multifunctionality:

1. The province may approve a "stepwise cleanup," which is a partial cleanup after which redevelopment may proceed before the final cleanup starts. Redevelopment may proceed after remediation of the most severe contamination to the Dutch List B-level. Redevelopment can not make A-level remediation impossible and remediation to the A-level must be part of the cleanup plan.

TABLE 10.1

RELATIVE COST OF CLEANUP THAT JUSTIFIES ON-SITE CONTAINMENT OF CONTAMINANTS (IN DUTCH GUILDERS, NLG)[a]

Cost of cleanup to Dutch List A-level is more than:	While cost of on-site containment is less than:	Relative cost factor
100,000,000	66,000,000	1.5
10,000,000	5,000,000	2
5,000,000	1,240,000	2.2
1,000,000	333,000	3
250,000	69,000	3.6
100,000	20,000	5
10,000	1,111	9

[a]Modified from The Netherlands Soil Protection Act of 1994.

2. If the environmental cleanup of the site to meet the A-level standard is technically impossible because of "technical location-specific circumstances," then the province may approve containment of the contaminants on-site.

3. If the remediation of the contamination will itself cause danger to public health, then the province may approve containment of the contaminants on-site.

4. If the environmental cleanup of the site to meet the A-level standard is too expensive, "financial location-specific circumstances" may apply. The province then compares permanent solution expense with the cost of on-site containment of contaminants and may approve containment on-site based on this comparison. The Soil Protection Act of 1994 clearly specifies what is "too expensive" (Table 10.1).

REMEDIATION IN THE NETHERLANDS

The Netherlands uses a full range of remediation techniques, the most common being excavation and removal to a landfill. Research indicates that 40–60% of remedial actions in The Netherlands involve removal and disposal in a landfill. The cost of landfilling contaminated soil ranges from 50 to 100 NLG per ton (P. van Arnheim, facsimile, April 23, 1994). Landfilling of contaminated soil is likely to be used less often in the future for two reasons: (1) the cost of landfilling is rising because of environmental requirements on the operation of landfills and (2) the Minister of the Environment is likely to prohibit the disposal of contaminated soil in landfills if it is technically possible to clean the soil (P. van Arnheim, facsimile, April 23, 1994).

The Netherlands has been a world leader in developing bioremediation techniques. Excavation and removal off-site to a location using bioremediation techniques has been the most common soil cleaning technique, occurring in approximately 85% of those cases (P. van Arnheim, facsimile, April 23, 1994). Incineration of soil contaminated with toxic and hazardous substances and *in situ* bioremediation on the contaminated site are also common.

Lekkerkerk Case Study

In The Netherlands, publicity about a contaminated site in the village of Lekkerkerk played an important role in the development of both public support and government policy toward contamination and environmental cleanup. The events in Lekkerkerk have numerous similarities to the case of Love Canal in the United States (see Chapter 1). For more complete information concerning the events at Lekkerkerk, refer to Hoomans and Stellingwerff (1982) or Kingsbury and Bingham (1992), the sources of the present discussion.

Lekkerkerk is a town in western Holland, located about 16 kilometers (10 miles) north of Rotterdam. In the 1970s, developers built a residential neighborhood of 268 houses on 8.9 hectares (22 acres) of land in an area below sea level and with high groundwater. Before construction, the land was "improved" by raising its level about 1 meter and filling in drainage ditches using waste building materials as fill. Unfortunately, the fill contained toxic wastes. Investigators identified the building industry, paints and varnish manufacture, paint spraying, plastics manufacture and processing, the chemical industry, and printing ink manufacture and application as the sources of the toxic contaminants.

They found several potential routes of toxic chemical exposure to humans. Heavy metals and organic contaminants, especially aromatic hydrocarbons, alcohols, ketones, and esters, were in the surface water, soil, and groundwater of the neighborhood. Contamination also existed in soil just below the floor of houses, and air samples from residential building crawl spaces had concentrations of toluene and xylene as high as 1000 ppm. They also found that these chemicals diffused into the low-density polyethylene (LDPE) water pipes in the neighborhood.

In the spring of 1980, the government evacuated the inhabitants and began remedial actions, which included removing contaminated soil. Workers cleared the ditches and nearby soils, and excavated the soil in the entire neighborhood to a depth of 0.7 meter (2.3 feet), and down to 3 meters (10 feet) under some houses. Technicians analyzed the soil to separate the clean from the contaminated. The government had not yet developed the Dutch List A-, B-, and C-values to use as a guide for determining contaminated soil. About 153,000 metric tonnes (168,654 tons) of contaminated soil were sent to an incinerator. Workers demolished a school and gymnasium on the site, and collected and treated surface water and groundwater before discharging it to the River Lek. The cost of the remedial actions at Lekkerkerk was 150 million NLG ($65 million U.S.). After remedial action, this residential neighborhood in Lekkerkerk was returned to its former use.

Tilburg Gas Works Case Study

Gas works operating more than 100 years ago in many European cities produced contaminated sites that are serious problems today. Redevelopment projects in European urban centers sometimes discover contaminated sites from gas works, tanneries, printing presses, or other activities that released persistent contaminants to the land. Dutch officials discovered the Tilburg Gas Works site when evaluating a proposal to redevelop an urban site into multifamily housing in central Tilburg, a major city in the province of North-Brabant. More complete discussion of this case is found in Kingsbury and Bingham (1992), which provided the information presented here.

The Tilburg Gas Works converted coal into gas for street lights and other uses for over 100 years until it closed and the structures were demolished in 1960. The facility operated on a 5.5-hectare (13-acre) site near the center of the city of Tilburg. The plant produced waste by-products of tar, benzene, toluene, naphthalene, and ammonia and disposed of production wastes on the site as fill material. There were also frequent spills and leaks.

In 1982, the city of Tilburg commissioned an environmental assessment of the Tilburg Gas works site. Investigators found soil and groundwater contaminated with volatile aromatic hydrocarbons, polynuclear aromatic hydrocarbons, and cyanides, the latter at concentrations up to 7700 ppm. They also found soil contaminated with tar and oil products over large areas to a depth of 4 meters and aromatics to a depth of 6–9 meters. High concentrations of benzene were present in groundwater at depths to 20 meters.

The goal of the remedial action plan was that "future inhabitants of the site should not be confronted with any perceptible pollution from the former gas works" (Kingsbury and Bingham, 1992, p. 233). Workers removed all "sensory detectable pollution" to a depth of 2 meters and in some locations to greater depth. They used ongoing groundwater pumping and treatment to control exposure to contaminants lying more than 6 meters below the surface.

CONCLUSIONS

The Netherlands has been developing contamination and cleanup programs since 1981, and they are still working to make them more effective and efficient. The basic problem of environmental cleanup in The Netherlands, as in other countries, is that there are many contaminated sites and limited money available for cleanup. The Dutch approach to this problem is to establish a priority list of contaminated sites and then complete cleanups on the highest priority sites first. This list approach has similarities to the U.S. National Priorities List (see Chapter 8), but there are important differences.

The Netherlands uses factors in addition to the environmental risk of a contaminated site when establishing its priority list. The risks to human health and ecosystems are important factors, but the redevelopment of contaminated brownfield sites for productive uses is also an important consideration, hence urban sites are at the top of the list and rural sites are at the bottom. The Dutch place strong emphasis on the potential future land use of the site, and municipalities have considerable input into establishing the priority of cleanups within their borders.

Yet some aspects of the contamination and cleanup problem have not been specifically addressed. For example, although The Netherlands policy is to clean up contaminated sites that are in use, they tend to overlook contaminated derelict land. Dutch policies do not allow municipalities to combine environmental cleanup budgets with housing or infrastructure budgets. This means that sometimes municipalities develop housing outside the urban boundary when they would have preferred to remediate a contaminated urban site and then develop it for housing. Such actions are often contrary to the Compact City Policy and the wishes of the municipality. To address this problem, the Dutch government has increased provincial budgets by 25 million NLG to remediate contaminated sites. This initiative is an attempt to eliminate contaminated sites that are "urban bottlenecks" in the construction of housing or other urban development projects (Business Roundtable, 1993, p. 83). This policy does not address orphan sites outside these urban bottlenecks.

The Netherlands has not been successful in recovering cleanup costs from responsible parties. During the period 1988–1991, the government attempted to recover expenses of 37 million NLG but recovered only 4.6 million NLG (Business Roundtable, 1993, p. 85). The Soil Protection Act of 1994 gives the government additional powers to recover expenses from polluting parties, and so the collection of such costs will likely improve.

References

Bernstein, J. D. 1993. *Alternative Approaches to Pollution Control and Waste Management: Regulatory and Economic Instruments.* Washington, DC: World Bank for the UNDP/UNCHS/World Bank Urban Management Programme.

Business Roundtable. 1993. *Comparison of Superfund with Programs in Other Countries.* Washington, DC: Business Roundtable, 117 pp.

Hoomans, J. P., and J. W. Stellingwerff. 1982. *Operation Lekkerkerk West.* Reprint from Pt/Civiele Techniek, 1982, No. 1, pp. 3–44.

Kingsbury, G. and T. Bingham. 1992. *Reclamation and Redevelopment of Contaminated Land: Vol. II: European Case Studies* (EPA 600/R-92–031). Cincinnati, OH: U.S. Environmental Protection Agency, Office of Research and Development, Risk Reduction Engineering Laboratory.

Ministry of Housing, Physical Planning, and the Environment. 1983. *Leidraad Bodembescherming (Soil Cleanup Guidance Note: Assessing the Severity of a Case of Soil Contamination in The Netherlands.* The Hague, April.

Needham D. B., B. Kruit, and P. Koenders. 1993. *Urban Land Property Markets in The Netherlands,* London: UCL Press, 228 pp.

Soczó, E., T. A. Meeder, and C. W. Versluijs. 1992. Ten years of soil cleanup in The Netherlands. Contribution to the Tour-de-Table. *Summary Report.* The 1992 NATO/CCMS Pilot Study Meeting on Evaluation of Demonstrated and Emerging Technologies for the Treatment and Clean-up of Contaminated Land and Groundwater, Budapest, October.

van Arnheim, P. 1994. *Dossier of Contaminated Property: Transaction, Lease, Development, Finance, Appraisal.* Amsterdam: V.T. Productions.

Visser, W. J. F. 1993. *Contaminated Land Policies in Some Industrialized Countries.* Report to the Technical Soil Protection Committee, The Hague.

CHAPTER 11

CONTAMINATED SITES AND ENVIRONMENTAL CLEANUP IN THE UNITED KINGDOM

INTRODUCTION

The United Kingdom has followed a distinctive policy approach toward contaminated sites. It has concentrated on reclaiming and redeveloping derelict land in urban areas, some of which is contaminated. The British approach coordinates urban and environmental policy to prevent urban blight and suburban sprawl and to restrict development in rural and urban fringe areas. Together these policies encourage the reuse of urban brownfield sites. In the past, the derelict land policy sometimes functioned to remediate toxic contamination. Recently, however, policy has shifted toward a greater focus on the cleanup of contaminated sites, yet has retained an emphasis on the future productive uses of such sites.

ORGANIZATION AND HISTORY OF CLEANUP EFFORTS

Within the United Kingdom, there is considerable decentralization of authority for contending with contaminated sites. Different offices have responsibility in different administrative units: the Department of the Environment in England, The Welsh Office in Wales, the Scottish Development Department (later the Scottish Enterprise) in Scotland, and the Department of the Environment for Northern Ireland in Northern Ireland. Under the Environmental Protection Act, local authorities are responsible for investigating contamination and taking remedial action when it is necessary. This chapter concentrates on England's Department of the Environment programs as illustrative of the United Kingdom approach.

Two separate approaches eventually evolved to form a coherent policy on contaminated sites. The United Kingdom developed one policy approach to the remediation and redevelopment of a limited number of seriously contaminated sites such as old landfill sites and former gas works. This approach attempted to find reasonable cost solutions (Saunders, 1987). The second policy approach concentrated on redeveloping derelict land, some of which was also contaminated.

The second approach focused on the value of contaminated land: "The legacy of dereliction arising from past industrial and extractive activity represents a significant waste of a vital national resource" (Department of the Environment, 1991, Part 1). Indeed, the definition of contaminated land reflects this focus on value; "Land so damaged by industrial or other development that it is incapable of beneficial use without treatment" (Department of the Environment, 1991, Part 18). In the United Kingdom, government policy is designed to encourage

> full and effective use of land in urban areas and the reuse of sites which have previously been developed. Recycling of land helps to revitalize urban areas and reduces the need to use new sites outside built-up areas, thus assisting the protection of the Green Belt and safeguarding the countryside. The re-use of contaminated land can contribute towards these objectives. (Department of the Environment, 1987)

Several initiatives are aimed at reclaiming and redeveloping derelict and contaminated land, among them the Urban Programme, Derelict Land Grant, Land Registers, Urban Development Grant, and Urban Regeneration Grant (the government combined some of these programs in a Single Regeneration Budget and renamed others). Urban Development Corporations are public–private partnerships that operate in designated districts (e.g., London Dockyards, Merseyside) and are active in reclamation and redevelopment within their project areas. The Water Resources Act gives the government the right to undertake remedial actions to clean up surface water or groundwater.

Building regulations in the United Kingdom also address the problem of contaminated land. Even if a local planning authority grants permission for a development, building regulations can force the cleanup of any contamination that is present. The regulations identify a number of toxic substances that require removal (Langham, 1993).

INTENDED FUTURE LAND USE

In all policies that affect contaminated sites, the United Kingdom has always considered the intended future use. The British use the phrases "suitable for use," "suitable for use as," and "fitness for purpose" to describe this approach. If required, remediation should control any unacceptable risks to health or the environment, taking into account the actual or intended use of the site.

The impetus to remediate contaminated sites has always been to redevelop the land and make it productive, thereby overcoming blight and supporting economic regeneration. This reclamation and redevelopment of contaminated land redirects development pressure away from greenfield sites, which helps prevent the urbanization of the countryside. This policy approach is in contrast to the " multifunctionality" approach used in The Netherlands, the United States, and other countries, in which remediation must render the land sufficiently safe for any future use. A British spokesperson explained the government's logic in rejecting the multifunctionality approach in this way:

> If disproportionate or unnecessarily early steps to treat land have to be taken, this may positively jeopardize wealth creation, by imposing excessive burdens on industry and commerce. It is necessary, therefore, to guard against over-specifying the degree of restoration, by considering the costs and benefits of particular schemes, and balancing the need for economic growth and environmental protection. (*Planning,* 1994)

In the United Kingdom, land use planning plays an important role in addressing contamination. One purpose of planning is to protect the human environment and one of the ways to achieve that goal is by regulating the use of land. The United Kingdom has a three-tiered planning system: (1) The Department of the Environment issues planning legislation and guidance (e.g., Planning Acts, Planning Policy Guidance Notes, Circulars, etc.), which provide the backbone of the land use planning system; (2) the national government requires county councils to produce Structure Plans that set out broad land use policies and the main proposals for the county; (3) the national government requires district councils to produce detailed development plans (Local Plans) that conform with the Structure Plan and also with national planning guidance. Where unitary authorities exist (e.g., London, metropolitan counties), they are responsible for drawing up a Unitary Development Plan, which consists of two parts: the first is similar to a Structure Plan and the second to a Local Plan.

Government advice states that land use plans should address the problem of contaminated land:

> In preparing their development plans, planning authorities need to take into account the possible effects on health and the environment of contaminated land. Development plans provide an opportunity to set out policies for the reclamation and possible use of contaminated lands. (Department of the Environment, 1994b, paragraph 4)

> [Local plans] should include detailed criteria which will be applied in determining planning applications for development on land which is known to be, or may be, contaminated. They may also set out any site-specific proposals for land use, where contamination is known or the site history suggests a risk of contamination, so that they may be readily identifiable to landowners and prospective purchasers or developers. (Department of the Environment, 1994b, paragraph 5)

Thus, the land use planning process plays a central role in cleaning up existing contaminated land, and to a lesser extent in preventing future contaminated land through regulation of future land use. When developing plans and reviewing specific development proposals, local planning authorities focus on issues of land use and leave the control of pollution to the pollution control authorities.

PROCESS FOR RECLAIMING CONTAMINATED SITES

The Interdepartmental Committee on the Redevelopment of Contaminated Land (1983) recommends a five-step process for the reclamation of derelict and contaminated land: (1) identification, (2) investigation, (3) assessment, (4) remedial action, and (5) monitoring. The process begins when a developer expresses interest in the redevelopment of a site. The proposed end use of the site determines the level of cleanup.

Under the UK approach, developers lead the entire five-step reclamation process. They initiate the process by expressing interest in potentially redeveloping a site (step 1). The local authority may inform the developer that the site is contaminated or that the site is likely to be contaminated because of previous land use. The developer then carries out a site investigation (step 2), followed by an assessment (step 3). Developers often conduct steps 1, 2, and 3 in consultation with a local authority, but they complete each step. On the basis of the assessment, the developer proposes remedial actions (step 4) sufficient for the intended land use, which they select subject to the existing land use plans. The monitoring of landfill sites is common in the United Kingdom, yet the government normally does not require monitoring of contaminated reclamation sites (step 5). This five-step process is still the norm in the United Kingdom, although it will likely change in the near future with a new statutory framework for identifying and dealing with contaminated land that is proposed as part of the new Environment Bill.

Step 4 of the reclamation process requires developers to propose and local authorities to approve remedial action plans. These plans determine what concentration of contaminants will be acceptable for the intended land use after the cleanup. The developer is thus responsible for determining whether the land, before or after cleanup, is suitable for the proposed development (Interdepartmental Committee, 1987, Annex Part 3). The remedial action plan also determines which remediation techniques will be used.

Guidance to the important question, "How clean is clean?" is not clear in the United Kingdom. Local authorities make this technical and often contentious decision when they approve or refuse the remedial action plans proposed by the developer. There are no legally binding action levels for concentrations of contaminants, but there are "trigger values" for 17 priority contaminants (Interdepartmental Committee, 1983).

Trigger values are concentrations of contaminants on land, and they accomplish two purposes: a threshold trigger value below which there is no significant risk of contamination and (2) an action trigger value at which the government determines that contamination is so serious that local authorities must require remedial action. Concentrations between the two values require site-specific judgment based on the present or intended future use of the site (Table 11.1). The first set of trigger values is based primarily on human health effects, but does not consider threats to ecosystems (Visser, 1993, p. 15).

The trigger values are helpful, but contamination is often complex and may involve numerous substances. The government must consider greatly expanding the number of substances for which there are trigger values, because the relatively small number of values provides little guidance when considering the thousands of possible toxic contaminants. There are no trigger values for contaminants in surface water or groundwater, which commonly occurs on or under a contaminated site. Nor are there action level trigger values for inorganic contaminants. The government expanded the list of trigger values in 1990 and is currently developing action levels. The list of trigger values is still inadequate.

The local authority reviews the investigation results, the assessment results, and the remedial action plan. Unfortunately, local authorities often lack the expertise to critically evaluate these technical results; however, they may consult the Interdepartmental Committee on the Redevelopment

TABLE 11.1

TRIGGER VALUES FOR CONTAMINATED LAND IN THE UNITED KINGDOM[a]

Substance	Planned use of site	Trigger value (mg/kg in air-dried soil) Threshold	Action
Selected Inorganic Contaminants			
Arsenic	Domestic gardens, allotments	10	TBD[b]
	Parks, playing fields, open space	40	TBD
Cadmium	Domestic gardens, allotments	3	TBD
	Parks, playing fields, open space	15	TBD
Chromium (VI)	Domestic gardens, allotments	25	TBD
	Parks, playing fields, open space		
Chromium (total)	Domestic gardens, allotments	600	TBD
	Parks, playing fields, open space	1,000	TBD
Lead	Domestic gardens, allotments	500	TBD
	Parks, playing fields, open space	2,000	TBD
Mercury	Domestic gardens, allotments	1	TBD
	Parks, playing fields, open space	20	TBD
Selenium	Domestic gardens, allotments	3	TBD
	Parks, playing fields, open space	6	TBD
Boron (soluble)	Any uses where plants are grown	3	TBD
Copper	Any uses where plants are grown	70	TBD
Nickel	Any uses where plants are grown	70	TBD
Zinc	Any uses where plants are grown	300	TBD
Contaminants Associated with Former Coal Carbonization Sites			
Polyaromatics	Domestic gardens, allotments, play areas	50	500
	Landscapes, buildings, hardcovers	1,000	10,000
Phenols	Domestic gardens, allotments	5	200
	Landscapes, buildings, hardcovers	5	1,000
Cyanide	Domestic gardens, allotments, landscapes	50	500
Free complex	Buildings, hardcovers	100	500
	Domestic gardens, allotments	250	1,000
	Landscapes	250	5,000
	Buildings, hardcovers	250	None
Thiocyanate	All proposed uses	50	None
Sulfate	Domestic gardens, allotments, landscapes	2,000	10,000
	Buildings	2,000	50,000
	Hardcovers	2,000	None
Sulfide	All proposed uses	250	1,000
Sulfur	All proposed uses	5,000	20,000
Acidity	Domestic gardens, allotments, landscapes	pH 5	pH 3
	Buildings, hardcovers	None	None

[a]Modified from Visser (1993) and Interdepartmental Committee (1987).
[b]To be developed.

of Contaminated Land. Nevertheless, conflicts of opinion between developers and planning authorities on technical issues of contamination, environmental health, and remediation are inevitable and developers may appeal against a refusal of planning permission.

Having developers manage the reclamation process for contaminated land has both advantages and disadvantages. In theory, it is economically more efficient to have the private sector conduct the investigation and assessment because market forces impose discipline to keep costs at a minimum. It is in the developer's interest to conduct a thorough investigation to have the best possible information on which to base investment decisions. Knowledge of the extent of contamination is critical when making decisions about the value of land and the cost of developing a site.

In practice, having private parties conduct the environmental appraisal used in cleanup decisions has serious disadvantages. The most obvious is the possibility of biased information. Where there is potential gain at stake, the appraisal may fail to provide the complete and unbiased evaluation that is best for the decision-making process. The owner of a parcel of contaminated land who wishes to sell it has a clear financial incentive to conduct a site investigation (environmental appraisal) that underestimates the extent of contamination. If such an investigation leads to local authority planning approval for a proposed development, that can greatly increase the value of the land and the ease of selling it.

Giving the developer responsibility for the cleanup of contaminated land contributes to another serious weakness of the UK approach. Namely, if there is no proposal to redevelop, there is little prospect that any remediation will take place except in the most severely contaminated cases where there is a clear and obvious threat to public health. Such a system driven by private sector development is in stark contrast to one based on the protection of public health.

Early policies in the United Kingdom backed into the problem of cleaning up contaminated sites through a concern to redevelop derelict land and revitalize urban areas. Until recently, this is opposite to the evolution of policy in the United States, where cleaning up contaminated land was the initial approach, and then policy evolved into concern for the future use of the land as a way to control the cost of cleanup. In May 1995, the U.S. EPA issued a directive that remedial action objectives developed during the Remedial Investigation and Feasibility Study should reflect the reasonably anticipated future land use or uses (U.S. Environmental Protection Agency, 1995).

In the United Kingdom, for severely contaminated land that is a clear threat to public health, local authorities can move quickly by serving notice on the "responsible party" and requiring them to remediate the contamination. If government cannot find the responsible party, then the local authority can require the owner or operator to remediate the contamination. If parties fail to act (a criminal offense), the local authority has the power to remediate the contamination and attempt to recover the costs from the responsible party through legal action. This process is quite similar to that in the United States.

Liability issues related to contaminated land are still evolving in the United Kingdom. There is strict liability for responsible parties, but the actual implementation concerning strict liability is not yet clear. In seeking cost recovery from current owners, local authorities must consider the financial hardship that this may cause (Business Roundtable, 1993, p. 96). Liability is not retroactive, which means that parties who released toxic substances that caused contamination before this activity was illegal are not liable. Innocent owners are theoretically liable, but the reference to hardship suggests that they may not be held liable. Directors of companies or real estate agents may be personally liable if they are guilty of neglect, consent, or connivance in relation to contaminated land (Business Roundtable, 1993, p. 98). The United Kingdom has not yet determined the liability of banks and local governments who may gain ownership of contaminated land through nonpayment of loans or property taxes.

DERELICT LAND GRANT PROGRAMME

The Derelict Land Grant Programme is the government's main program to reclaim derelict land and has been in operation the longest. The program may also allow local government to help deal with contaminated land in cases where it is not feasible to make the polluter pay (Department of the Environment, 1994c). The program's budget has been substantial for more than a decade (Table 11.2).

Note that Table 11.2 shows the budget of only one program, and that in inner city areas the Urban Development Corporations, City Grants, and the Urban Programme (now the Single Regeneration Budget) also reclaimed much derelict land. The total UK government expenditure in 1991 was between £150 million and £200 million per year (Department of the Environment, 1991. Thus the Derelict Land Grant Programme represented approximately half of government expenditures on derelict land.

This program has reclaimed a modest quantity of land that has remained relatively steady, near the 1979–1989 average value of 1314 hectares (3246 acres) per year (Table 11.3). Though the total amount of reclaimed land is not large, it contributed to achieving the policy objective of putting this "vital national resource" back into productive use, rather than

TABLE 11.2

DERELICT LAND GRANT PROGRAMME RESOURCES SINCE 1980[a]

Year	Amount (millions of £ in 1991 constant prices)
1980–81	57.4
1981–82	66.3
1982–83	126.4
1983–84	118.0
1984–85	112.7
1985–86	118.0
1986–87	114.5
1987–88	105.7
1988–89	93.9
1989–90	76.7
1990–91	75.8
1991–92	87.9
Total	1153.3

[a]From Department of the Environment (1991).

wasting it by allowing it to remain unproductive. In fact, in 1986, "almost half of all new development was on previously used land" (Department of the Environment, 1987, paragraph 41061).

These data provide a rough estimate of the cost of reclaiming derelict land in the United Kingdom. When we compare the 11 years of data for land reclaimed to the first 11 years of Derelict Land Grant Programme costs, we find an approximate average cost of £73,700 per hectare ($45,950 U.S. per acre). Note that this is only a rough estimate and that the assumptions are not necessarily valid. Some of the land that the Department of the Environment reclaimed using the Derelict Land Grant may have been reclaimed even without the grant.

Information on how much of the reclaimed derelict land was contaminated is not available. The cost of reclaiming contaminated land is usually much greater than the cost of reclaiming derelict but uncontaminated land. We should assume that much of the reclaimed land represented in Tables 11.3, 11.4, and 11.5 was derelict land, but not contaminated by toxic chemicals. The presence of toxic contamination causes great variation in the reclamation cost.

Local authorities conducted derelict land surveys for the Department of the Environment in 1974, 1982, 1988, and 1991. Table 11.4 reports the

TABLE 11.3

DERELICT LAND RECLAIMED SINCE 1979[a]

Year	Amount (hectares)
1979–80	1,236
1980–81	1,403
1981–82	1,742
1982–83	1,444
1983–84	1,364
1984–85	1,343
1985–86	1,060
1986–87	903
1987–88	1,285
1988–89	1,487
1989–90	1,183
Total hectares	14,450
(Total acres)	(35,706)

[a]From Department of the Environment (1991).

data from the 1982 and 1988 surveys. During this period, derelict land declined by 5200 hectares (12,849 acres) or 11%. The reduction was 4% in urban areas and 17% in rural areas. Table 11.4 also reports the amount of derelict land that the local authorities, who completed the surveys, thought justified reclamation. In their opinion, derelict urban land justified reclamation to a substantially greater extent (92–95%) than rural derelict land (61–62%).

TABLE 11.4

CHANGE IN THE AMOUNT OF DERELICT LAND IN URBAN AND RURAL AREAS, 1982–1988 (IN THOUSANDS OF HECTARES)[a]

	Urban		Rural		Total		Percent change 1982–1988
	1982	1988	1982	1988	1982	1988	
Total derelict land	21.0	20.1	24.7	20.4	45.7	40.5	−11
Area justifying reclamation	19.3	19.0	15.0	12.6	34.3	31.6	−8

[a]From Department of the Environment (1991).

TABLE 11.6

DERELICT LAND GRANT (DLG) EXPENDITURE PRIORITIES, 1985–1990, IN PERCENTAGE OF LAND RECLAIMED[a]

	1985–1986	1986–1987	1987–1988	1988–1989	1989–1990
DLG for:					
Hard end use	56	42	46	61	52
Soft end use	44	58	54	39	48
DLG expenditure in inner city areas	30	32	39	40	40

[a]From Department of the Environment (1991).

Overall, the Derelict Land Grant Programme has had a substantial effect in encouraging reclamation of derelict land. Table 11.5 indicates that local authorities using program grants reclaimed more than half of the derelict land reclaimed in the United Kingdom in the years 1982–1988. Yet the quantity of derelict and contaminated land is not static. Even as the program is affecting the reclamation and reuse of existing derelict land, new lands are becoming derelict.

Up to 1991, the Derelict Land Grant policy was to direct grants to the inner cities to support the government's initiative to regenerate and revitalize them. This policy gave priority to "hard end uses" that have a higher propensity to attract private sector investment to reclaimed land and to produce significant economic benefits to the cities. Hard end use usually means that a developer constructs a facility on the remediated site. Alternatively, a soft end use might be a park. Table 11.6 shows that between 1985 and 1990, between 42 and 61% of Derelict Land Grant Programme expenditures went to hard end uses. It also reveals that the percentage of expenditures going to inner city areas increased from 30 to 40% over this same period.

Beginning in 1991, however, the Department of the Environment changed its policy toward derelict land to make it more strategic in focus and more proactive. This new strategic focus recognized that reclamation schemes are most effective when they are an integral part of a local regeneration strategy (Department of the Environment, 1991, paragraph 19). This approach developed from experience with "rolling programs."

Local authorities use rolling programs to tackle the most extensive areas of dereliction in a comprehensive manner rather than using a project-by-project approach to reclamation. An essential aspect is guaranteed

TABLE 11.7

AFTER VALUE FROM DERELICT LAND PAID TO THE DEPARTMENT OF THE ENVIRONMENT[a]

Year	Amount (millions of t)
1986–87	2.07
1987–88	2.26
1988–89	10.35
1989–90	11.44
1990–91	12.42
1991–92[b]	12.35

[a]From Department of the Environment (1991).
[b]Forecast.

funding over a number of years. In 1991, 21 rolling programs were in operation, and both internal Department of the Environment and external audits judged them to be successful (Department of the Environment, 1991, Part 20). The Department has published guidance to local planning authorities on how to follow this strategic approach (Johnson *et al.*, 1992).

Derelict Land Grant Advice Note 1 of 1991 also informed local authorities that the Department of the Environment was establishing a special program that would allow Supplementary Credit Approvals for landfill gas works not associated with the reclamation scheme and thus not eligible for the Derelict Land Grant Programme. These credit approvals are available only for remedial works at closed landfill sites and will not be approved for a site in the Derelict Land Grant Programme.

The other significant aspect of the 1991 changes was to make the policy more proactive. To this end, the Department of the Environment instructed local authorities, under certain circumstances, to take action to remediate contamination before a developer proposes any redevelopment project that would require remediation. Furthermore, it placed greater emphasis on the option of reclaiming derelict (and contaminated) land for soft end uses such as parks and open space. Local planning authorities were also given greater flexibility to use grants for amenity use or environmental improvement as opposed to reclaiming for hard end uses, which had prevailed in the late 1980s.

Thus the 1991 policy guidance allowed more flexibility in selecting projects as long as they conformed to locally developed reclamation strategies. Reclamation requires this flexibility because derelict land exists in both urban and rural areas, and because reclamation to a hard end use standard is expensive. This new policy direction aspired to achieve several disparate objectives, including (1) treating the worst dereliction first, (2) improving the environment for recreation, nature conservation, and historic conservation, and (3) tree planting to help establish community forests near urban areas. In the words of the Department: "Priority within the Derelict Land Grant Programme will be given to the treatment of land which in its present condition reduces the attractiveness of an area as a place in which to live, work, or invest, or because of contamination or other reasons is a threat to public health and safety or the natural environment" (Department of the Environment, 1991, paragraph 16).

In the United Kingdom, the government captures some of the value produced by the reclamation process to help fund the Derelict Land Grant Programme, which in turn awards grants to local authorities. This increase in value is called "after value." The after value is the increased value of the property that the Derelict Land Grant Programme caused. When the government sells reclaimed land, the after value is returned to the program as part of its funding. This cumulative after value has increased over the years of the program (Table 11.7).

REGISTERS OF CONTAMINATED LAND

The Environmental Protection Act of 1990 required local authorities to compile and maintain registers of potentially contaminated land that would help establish the extent of the contamination problem as well as identify specific parcels of land. These local registers were to be based on "previous uses" of the land. Some suggested indicators of potential contamination were: manufacture of gas, coke, or bituminous material from coal; manufacture of lead, steel, or steel alloys; manufacture of petroleum or other chemicals; disposal of household, industrial, or commercial waste; treatment of waste by chemical or thermal means; and use of waste for scrap metal (Langham, 1993). The government intended the register to record whether remediation had occurred, but make no assessment of the degree or effectiveness of any cleanup.

Although the creation of registers of potentially contaminated land is a logical step with great utility, the process encountered political difficulties. Before long, the government abandoned the registers in light of three primary criticisms:

> (i) that since the proposed registers would be based on the criterion of use or former use by a limited range of potentially contaminating activities it would include a large number of sites that are not actually contaminated, while missing other sites that are actually contaminated by other former uses not on the prescribed list;
> (ii) that since the register would record past potentially contaminative uses of land there would be no way of removing such sites from the register even if any contamination had been satisfactorily dealt with;
> (iii) that when sites are identified as actually contaminated it remains unclear in some cases what action should be taken, what remediation measures should be carried out and by whom, which regulatory authorities should be involved, and where the liability for the cost of remediation or compensation should fall. (Howard, 1993)

Hence the creation of local registers was politically unacceptable. When local authorities listed a property on a register, the value of the property decreased. Landowners and the development community vigorously complained that this was unfair and that it depressed property markets. Some government officials believed that the registers potentially undermined land reclamation and urban regeneration efforts, thereby contributing to urban blight. Yet abandoning the registers was akin to blaming the messenger for the bad news. Actual contamination, or the information that suggested its presence, is a fact that exists with or without registers. Real estate markets that function properly should use this information whether a register exists or not.

In May 1993, the government announced a wide-ranging review of contaminated lands policy. More than a year later, this review produced a modified policy described in two documents (Department of the Environment, 1994a, b). The review restates the earlier development-oriented policy that remediation should meet the "suitable for use as" standard, which means that remediation should make the contaminated site safe for the intended future use of the property. The Derelict Land Grant Programme would continue, albeit in a modified form under the control of the newly created urban regeneration agency, English Partnerships. This entity, established on April 1, 1994, has the objective of "promoting job creation, inward investment and environmental improvement through the reclamation and development of vacant, derelict and under-used or contaminated land and buildings" (English Partnerships, 1994, p. 4).

Beginning in April 1995, English Partnerships assumed full responsibility for derelict land reclamation resources and was charged with returning derelict land to beneficial use. The new grant regime is a unified system of support for regeneration entitled the Investment Fund. The English Partnerships Land Reclamation Programme now operates this fund, which consists of former Derelict Land Grant and City Grant budgets, both devoted to dereliction. The government expects that, between April 1995 and April 1997, the level of funding and the methods by which it is allocated will remain similar to the former Derelict Land Grant regime. However, a flexible system of bidding will replace the fixed annual bidding cycles. In the longer term, the government expects that English Partnerships will formulate its own derelict land program, a unified grant system reflecting a greater emphasis on partnership working, leveraging private finance, and drawing on greater local community input.

CONCLUSIONS

Compared to other economically developed countries, the United Kingdom has been slow to establish effective policy to promote the cleanup of contaminated sites. In a sense, the United Kingdom "backed into" its current policy through concern about the economic revitalization of blighted urban areas. Despite this slow start, the government is finally developing contamination and cleanup policies, and is actively remediating contaminated sites. As the European Union develops a common policy for contaminated sites and cleanup, the United Kingdom may be forced to reevaluate its policies and bring them into line with those of its fellow EU countries.

Yet the United Kingdom's approach to environmental cleanup and the promotion of subsequent productive use may offer important lessons. Why should countries spend huge sums of money on contaminated sites that they will then leave derelict and susceptible to urban blight and socially unwanted activities? It seems logical that they attempt to gain further benefit by returning sites to beneficial use. The United Kingdom approach succeeds in this and is a model for other countries.

References

Business Roundtable. 1993. *Comparison of Superfund with Programs in Other Countries.* Washington, DC: Business Roundtable, 117 pp.

Department of the Environment. 1987. *Development of Contaminated Land,* Circular No. 21/87. London: HMSO. (Also Welsh Office Circular No. 22/87.)

Department of the Environment. 1991. *Derelict Land Grant Advice: Derelict Land Grant Policy* (DLGA 1). London: HMSO.

Department of the Environment. 1994a. *Paying for Our Past: The Arrangements for Controlling Contaminated Land and Meeting the Costs of Remedying the Damage to the Environment,* in Consultation with the Welsh Office. London: HMSO.

Department of the Environment. 1994b. *Planning Policy Guidance, Planning and Pollution Control, No. 23, Annex 10: Contaminated Land.* London: HMSO.

Department of the Environment. 1994c. *Assessment of the Effectiveness of Derelict Land Grant in Reclaiming Land for Development.* London: HMSO.

English Partnerships. 1994. *Investment Guide.* London: English Partnerships.

Howard, M. 1993. Written answer to a parliamentary question in the House of Commons, by the Secretary of State for the Environment, March 24.

Interdepartmental Committee on the Redevelopment of Contaminated Land. 1983. *Guidance on the Assessment and Redevelopment of Contaminated Land,* Note 59/83. London: HMSO.

Interdepartmental Committee on the Redevelopment of Contaminated Land. 1987. *Development of Contaminated Land,* DOE Circular 49/77. London: HMSO. (Also Welsh Office Circular 3/77).

Johnson, D., S. Martin, G. Pearce, and S. Simmonds. 1992. *The Strategic Approach to Derelict Land Reclamation.* Department of the Environment, Directorate of Planning Services. London: HMSO.

Langham, R. 1993. Contaminated land: The legal aspect. *J. Plann. Environ. Law* September, 807–815.

Planning. 1994. Ministers consider public duty to act over contaminated land. No. 1059, March 11.

Saunders, P. 1987. The UK approach to soil and landscape protection. In H. Barth and P. l'Hermite (eds.), *Scientific Basis for Soil Protection in the European Community.* London: Elsevier Applied Science.

U.S. Environmental Protection Agency. 1995. *Land Use in the CERCLA Remedy Selection Process.* OSWER Directive No. 9355. 7–04, Washington, DC.

Visser, W. J. F. 1993. *Contaminated Land Policies in Some Industrialized Countries.* Report to the Technical Soil Protection Committee, The Hague.

CHAPTER 12

Contamination and Environmental Cleanup in Central and Eastern Europe

INTRODUCTION

The countries of Central and Eastern Europe (CEE) have had serious difficulties in developing effective policies to prevent the creation of contaminated sites and to clean up existing contamination. In addition to the problems of other industrially developed countries, the nature of their post-World War II economic systems has caused others, such as floundering, inefficient economies and the necessity for political and economic transitions.

The state of environmental law varies considerably among the CEE countries. Some have developed contemporary legislation capable of contending with their problems, whereas others still rely on old laws established under communist regimes. CEE countries have a history of civil law that imposes damages only if a firm violates a government standard or regulation. The concept of liability for the cost of environmental cleanup is unclear in these legal systems. However, there is considerable movement toward the "polluter pays principle," but CEE countries will need appropriate laws and a well-implemented enforcement scheme to institutionalize them if they are to make progress in controlling environmental problems.

Within Central and Eastern Europe, there are two distinct groups of countries in terms of their progress toward achieving environmental cleanup. This distinction is primarily a function of their progress in the transition toward democracy and market economies. One group consists of the

Central European countries and the Baltic countries of Estonia, Latvia, and Lithuania. They have made greater progress in their economic transitions and are more likely to improve their environmental policy. Several of these countries have a goal of membership in the European Union.

The second group of countries are those of the former Soviet Union, excluding the Baltic countries. This is known as the Commonwealth of Independent States (CIS). The CIS states have made less progress in their political and economic transitions and are less prepared than the countries of Central Europe to undertake cleanup of contaminated sites.

ENVIRONMENTAL DAMAGES IN CENTRAL AND EASTERN EUROPE

The centrally planned economic systems of the postwar CEE countries left environmental damages that are the worst on the planet. Their previous economic systems contributed in several ways (Wilczynski, 1990). Such economies concentrated on heavy industry, lacked advanced pollution control technologies, and had incentives to increase production at any cost. CEE countries also lacked a conservation ethic and their governments placed restrictions on environmental interest groups. As a result, contaminated brownfield sites are among the worst problems in many of these countries.

Contaminated property, especially the sites of former industrial facilities in urban areas, are among the most visible environmental problems in the CEE countries. Military, mining, and energy production activities also produced great quantities of contaminated land. These environmental hot spots receive a great deal of publicity, but little environmental cleanup is taking place. Yet even though contamination problems are widespread, they are probably not the most serious environmental problem that these countries face.

The CEE countries must deal with the legacy of this old environmental degradation, but at the same time must confront the new damages resulting from changes associated with the political and economic transition to democratic governments and market economies. Market reforms have set free a strong entrepreneurial spirit, and many new, aggressive businesses have appeared and operate in a largely unregulated fashion. Maximization of profits has become all-important, but without sufficient controls to protect the environment and other public values.

Though evidence is scarce, there are widespread reports of unscrupulous entrepreneurs shipping hazardous wastes from other industrialized countries and disposing of them in CEE countries (Clark, 1995). This is done despite treaties and laws prohibiting transboundary shipping of hazardous wastes.

Furthermore, as production costs increase, competition intensifies, and government subsidies disappear, there are strong incentives for industry and business to reduce costs. Reducing the expense of environmental protection is sometimes an attractive option in the CEE countries, where government enforcement of environmental programs is weak. Overall economic conditions in the near term also present a serious obstacle to the cleanup and reuse of contaminated sites. Rising unemployment, falling real wages, and high inflation have made economic development the predominant concern of the CEE population. Citizens are more concerned with price increases and job security than with environmental protection.

Economic difficulties over the past decade or more in the CEE countries have also resulted in insufficient maintenance of the industrial infrastructure. For example, in Russia, 10% of the 550,000 kilometers of oil pipeline has been in use for more than 35 years. Experts consider pipes in use for 10 years to be "risky" (Wesolowsky, 1994). Indeed, in a single incident in 1994, over 300,000 tons of oil spilled in the Komi region of Russia, which has been nominated as a World Heritage Site (*Environmental Cooperation Bulletin,* 1994a). This region is one of the last, large virgin stands of coniferous forest left in the world. Pipeline leaks are common and will continue to create additional contaminated sites until officials implement appropriate maintenance practices.

Yet the transition to democratic governments and market economies has produced some environmental benefits. It has decreased the output of heavy industry, due to competition and a reduction in government subsidies, which has reduced pollution discharges by about 20–30%. There has been some restructuring of industrial sectors that has led to retrofitting existing plants and replacing obsolete technologies. The transition to a market economy has also created incentives for more efficient production practices and reduced waste discharges.

CASE STUDY OF CONTAMINATION AT FORMER SOVIET MILITARY BASES

Military operations create contaminated sites that require cleanup in all countries, but the CEE countries have especially serious problems in this area. Peacetime as well as wartime military operations cause contamination. Leaks, spills, and improper disposal of petroleum fuels are the most common forms of contamination, though toxic chemicals used in solvents, degreasers, and other common products are also common contaminants on military bases. An indication of the volume of waste produced is the U.S. military which produced 530,000 tons of hazardous wastes at 333 military installations in 1983 (Edelstein, 1988, p. 3). Munitions may also contaminate land at military bases and present special problems. Radioactive contamination is present at some military facilities.

The breakup of the former Soviet Union provides an interesting opportunity to learn something about contaminated sites on military bases. In most countries, information on such contamination is not available to the public. As the former Soviet army withdrew from the former Warsaw Pact countries, information on contamination at former Soviet military bases started to become available. Hungary was among the first countries to investigate contamination on its former Soviet military bases. Although a complete assessment of the contamination will take many years, initial information provides some insight to the problem (Reiniger, 1992).

On June 19, 1991, the last of the Soviet Union's 110,000 military personnel departed Hungary, but they left behind a staggering contamination and cleanup problem. The Soviet Army left approximately 500,000 tons of hazardous material spread over 48,000 hectares of land on seven bases with 6000 buildings (Embassy of the USA in Hungary, 1993). Most of the known contamination is from petroleum products, primarily aviation and land vehicle fuels, but munitions and other contamination are also present. Some of these contaminated sites are a serious threat to human populations. For example, some bases are located above karst aquifers near communities that draw their drinking water from these aquifers.

The Hungarian government is presently remediating some of this contamination and some redevelopment of formerly contaminated military bases is taking place. But Hungary has a limited budget for remediation and is in the middle of a painful economic transformation to a market-based economy. As a result, the primary goal of remediation is to contain the spread of contamination from the former military bases. There is little active cleanup to eliminate the contamination.

Most of the active remediation in Hungary is removing fuel from groundwater. Authorities can recover valuable hydrocarbons from the spilled and leaked fuels, which they can reuse or sell to produce revenues to offset some remediation costs. The former Soviet air force bases at Sarmellek and Tokol and the former Petfurdo petroleum fuel storage facility near Veszprem are sites where the government is capturing valuable hydrocarbon fuels from the ground.

However, there is no evidence that the government will continue remediation to clean up contaminated soil and groundwater once they can no longer cost-effectively recover valuable fuels. Where heavy metals contamination makes remediation more difficult, for instance, at the former

Lovasbereny tank base, where grease containing cadmium is one of the contaminants, cleanup is slow.

An example of an effective cleanup is the former Soviet heliport at Szekesfehervar. Kerosene spills and improper disposal caused serious contamination that threatened groundwater supplies used for drinking water by nearby communities. The government used biological stimulation technology to remediate the contaminated soil to the satisfaction of the local regulatory authorities (Embassy of the USA in Hungary, 1993). Under the privatization program, a private company bought this remediated former base to redevelop the property and exploit its location near the M7 highway.

The privatization of government-owned land, including former Soviet military bases, offers some prospects for remediation of contamination. Resolution of environmental liability for renters or new owners of contaminated property is necessary. The Hungarian government should also be willing to reduce the rent or purchase price of contaminated properties to induce their use or purchase. It is possible that such arrangements could encourage new owners or users of contaminated sites to remediate contamination and redevelop the property.

Other CEE countries have military contamination problems similar to those in Hungary. The Estonian Ministry of the Environment released a study of the amount of damage caused to the environment by 50 years of Soviet-Russian military occupation. A research team assessed the damage to surface water, groundwater, soil, and the man-made environment. They estimated the cost of repairing the damage at 58.95 billion EEK (approximately $4.3 billion U.S.). That total includes 54.8 billion EEK for direct costs and 4.4 billion EEK for nonpayment of environmental taxes (*Environmental Cooperation Bulletin,* 1994b). The cleanup of former Soviet military airfields and artillery ranges (costing over 11 billion EEK) and two areas with radioactivity represent the greatest cost. The two radioactive sites are the Paldiski nuclear submarine training facility (40 billion EEK) and the town of Sillamaee, which has radioactive waste from a uranium plant (1.7 billion EEK) (*Environmental Cooperation Bulletin,* 1994b). The research team has analyzed about 85% of the territory that was in the control of the Soviet military. Not included in the study was the evaluation of the damage caused to Estonian forests.

DEVELOPMENT AND ENVIRONMENTAL CLEANUP

The CEE countries have choices in how they can establish policies to promote the environmental cleanup of their contaminated sites. In making these choices, they can learn from the experience of the countries of Western Europe and North America. In this regard, the industrialized CEE countries face many of the same decisions as the less developed countries. These decisions include how to promote environmental protection and cleanup at the same time they are striving to promote economic development (see Chapter 7).

One strategy is to pursue economic development at almost any cost to achieve a level of prosperity that will generate the funds to pay for environmental cleanup. In this approach, countries recognize environmental needs but delay addressing them until their market economies have strengthened and can sustain sufficient economic growth. This approach considers economic growth to be a prerequisite for environmental improvements.

This strategy follows the history of industrialization in the market economies of Europe and North America, where environmental degradation eventually prompted extensive environmental investments. Some of the CEE countries seem ready to ignore environmental protection to realize economic recovery, even though this strategy is likely to exacerbate existing problems and create new ones. There is also a potential danger that economic factors may force CEE countries to specialize in industries that produce toxic substances.

The CEE countries can adopt a sustainable development approach as an alternative to following the historical path taken by the West. In this approach, countries combine economic development with environmental protection. Following this approach requires many hard decisions that some groups may view as restricting the growth of jobs and the generation of wealth. However, requiring industries to invest in environmental protection will force them to acquire modern, environmentally friendly technologies. Although this investment may be expensive in the short term, in the medium to long term it will be good for business because it will make the country's products and services competitive on world markets. It will also help avoid possible trade barriers. Aiming to achieve sustainable development may be essential to countries that hope to meet the membership requirements of supranational organizations such as the European Union.

WASTE DISPOSAL AND THE CREATION OF NEW CONTAMINATED SITES

The proper disposal of toxic substances is essential to avoid the creation of additional contaminated sites. Most of the CEE countries have not yet developed an effective infrastructure for hazardous wastes or for municipal solid waste disposal. Hazardous waste management has become a high-profile issue, with each country having its own hazardous waste problem. As mentioned earlier, serious problems are also caused by the movement of hazardous waste from abroad into the CEE countries for improper disposal.

ALTERNATIVE APPROACHES FOR ENVIRONMENTAL CLEANUP

As the CEE countries undergo fundamental political and economic changes, they have important choices to make concerning contaminated sites and environmental cleanup. How will these countries obtain the large sums of money needed to pay for the cleanups? This historic period of dramatic change provides a unique climate for introducing new approaches to achieve environmental improvements. The basic dilemma is how to allocate environmental cleanup costs between government and the current or future owners of contaminated sites.

There are advantages and disadvantages to either approach to generating the funds to pay for environmental cleanups. Public funding of cleanups seems appropriate because the Soviet bloc governments of the CEE countries are responsible for causing the toxic substances contamination of land and groundwater. Public funding produces a business climate that helps attract domestic and foreign investment because investors face reduced risk of being liable for the cost of an unexpected cleanup. Reducing this risk is important in the CEE countries, where there is a lack of established liability law, precedent, and consistent enforcement. Public funding can also produce a more orderly process that is able to prioritize contaminated sites and remediate the most dangerous sites first. The most serious disadvantage of public financing of environmental cleanups is that it does not provide an incentive to private firms to invest in pollution reduction. Consequently, this approach may encourage firms to increase their profits by not properly handling or disposing of toxic substances.

Use of an environmental liability approach to force private parties to pay for cleanup would provide an incentive to prevent future contamination, but this approach has several disadvantages. Private funding of cleanup might hinder privatization and industrial development because many contaminated industrial sites may have a negative value, with the cost of cleanup exceeding the value of the property. This would discourage foreign investment in CEE countries, which is important for bringing in skills, technology, and capital.

Some of the CEE countries have developed innovative systems for financing environmental projects. Several have instituted a State Fund for Environmental Protection to provide grants and loans for such projects (Bolan, 1992). These countries levy discharge fees and noncompliance fines, and also impose taxes and fines to generate cleanup funds. These funds can be highly effective when applied to direct environmental improvements, but in practice the allocation of such funds by CEE governments has been erratic.

APPROACHES TO ALLOCATING LIABILITY

The large number of contaminated sites and the high cost of cleanup mean that decisions about liability are important factors in the success of privatization and efforts to attract foreign investment. CEE governments would like to eliminate the uncertainty about the extent of liability and the cleanup costs for any privatization or sale of property. By eliminating uncertainty, potential investors can make informed decisions. The CEE countries have several mechanisms of allocating cleanup liability: to indemnify (wave the liability of) the purchaser, or to reduce the price of the property to a value that induces the purchaser to risk accepting liability (Page and Rabinowitz, 1994).

Governments that accept the responsibility to pay for cleanup must then devise means of generating the funds. Use of general revenues is possible, but during this transition period, most CEE countries have insufficient revenues and are operating at a deficit. Government funding would be a form of "pooled funds" that would operate like government-provided "no-fault" insurance. This approach offers the greatest incentives to privatization and foreign investment, but may encourage private firms to spend government funds in cleaning up insignificant contamination. Some analysts propose that public financing of cleanup through a pooled liability fund may be the best approach in the short term. In the longer term, however, countries need to develop comprehensive policies to fund and manage the remediation of contamination (Boyd, 1993).

Some CEE countries, for instance, Poland and the former Czechoslovakia, now The Czech Republic and the Republic of Slovakia, set up liability escrow accounts to pay for environmental cleanups. At the time of sale, governments place a proportion of the purchase price of a state-owned enterprise into this account. The purchaser can use these funds only to remediate contamination on the property that they purchase. If the purchase does not use the funds within one year, then the government may apply the funds to pay for the remediation.

CONCLUSIONS

The current period of dramatic and rapid change in the Central and Eastern European countries, as they undergo transitions in political and economic institutions, presents an opportunity for them to radically improve their environmental policies. Many CEE countries are privatizing state-owned enterprises to achieve the increased efficiency and productivity of market economies. Privatization, when combined with appropriate government remediation programs, can in turn lead to the cleanup of the purchased properties. This period of political and economic reform is also an opportunity to learn about and implement policies and remediation techniques used in other countries.

The CEE countries face serious problems in the economic, social, and political realms, as well as in the environmental arena. They are confronted with hard choices in how to equitably allocate their insufficient resources. It is unlikely that governments have the resources to perform cleanups on any but the most contaminated and health-threatening sites. Nonetheless, it is critical that they recognize the magnitude of this problem and begin developing effective environmental policies.

References

Bolan, R. 1992. *Organizing for sustainable growth in Poland. J. Am. Plann. Assoc.* **58**(3), 301–311.

Boyd, J. 1993. The allocation of environmental liabilities in Central and Eastern Europe. *Resources* Summer, No. 112, 1–6.

Clark, R. 1995. Australian toxic waste pollutes Russian environment. *Ecodefense! Inf.* No. 6(43), March.

Edelstein, M. 1988. *Contaminated Communities: The Social and Psychological Impacts of Residential Toxic Exposure.* Boulder, CO: Westview Press.

Embassy of the USA in Hungary. 1993. Former Soviet Bases: Some Progress but the Environmental Mess Remains. Cable from Environment, Science, and Technology Attaché to U.S. State Department, Budapest, May 7.

Environmental Cooperation Bulletin. 1994a. Vol. **3**(6), April.

Environmental Cooperation Bulletin. 1994b. Vol. **2**(12), October.

Page, G. W., and H. Rabinowitz. 1994. The potential for redevelopment of contaminated brownfield sites. *Econ. Dev. Q.* **8**(4), 353–363.

Reiniger, R. 1992. Overview of Hungary's environmental situation. *Summary Report.* The 1992 NATO/CCMS Pilot Study Meeting on Evaluation of Demonstrated and Emerging Technologies for the Treatment and Clean-up of Contaminated Land and Groundwater, Budapest, October.

Wesolowsky, T. 1994. Industrial accidents skyrocket. *Moscow Tribune* No. 150(363), August 9, p. 6.

Wilczynski, P. 1990. *Environmental Management in Centrally-Planned Non-market Economies of Eastern Europe,* Environment Working Paper No. 35. Washington DC: World Bank, 46 pp.

CHAPTER 13

Concluding Comments on Contamination and Environmental Cleanup Issues

INTRODUCTION

All countries have sites that are contaminated with toxic substances. The economically developed countries have many sites, with some countries having hundreds of thousands. The developing countries have fewer contaminated sites, depending on their history of industrial activity. However, the developing countries have a serious problem because many industries that use toxic substances are relocating there before the national governments can design and establish effective policies to prevent toxic pollution or remediate contaminated sites.

Contaminated sites represent a serious threat to the health of people working or living near them and to the ecosystem. Other sites are a potential threat, or the degree of risk may be unknown. Some contaminated sites have the effect of lowering the value of all properties within several miles. They may also negatively influence the prospects for economic development and growth in surrounding neighborhoods. Some sites act as magnets for socially unwanted activities and contribute to urban degeneration and blight. Almost all sites are a political problem because the public fears them and demands action to ameliorate health risks.

PUBLIC DEMAND FOR ENVIRONMENTAL CLEANUP POLICY

Fear of adverse health effects from exposure to toxic substances is the major motivation behind public outcry for government action. Though some contaminated sites do directly threaten public health, their number is small in comparison to the large number of contaminated sites that represent a *potential* health threat. Scientific evidence about the significance of the public health threat is inconclusive, and it is unlikely that well-founded conclusions will be reached in the near to medium term.

The public's demand for government action to clean up contaminated sites is strong, and this support exists irrespective of objective measurements of the human health risk. In most countries, the public demands thorough remediation regardless of risk assessments or benefit–cost analyses. Government efforts to improve risk communication seem unlikely to narrow the gap between public and expert perceptions of the risk of contaminated sites.

There is a strong fear that "toxic time bombs" may expose individuals and families to dangerous health threats in the present or the future. Actual health risks may exist without the public's knowledge because they may not see, smell, or taste the toxic substances threatening their health. This understandable fear leads to their demand that the government protect them from the risk of exposure to toxic substances, which may migrate from contaminated sites. Furthermore, the public may demand complete protection, to the extent that the remediated site represents zero risk. These perceptions of contaminated sites and public support for cleanup policies are likely to continue.

POLICY APPROACHES

In response to public demand for action, governments need three policy approaches. The first is to create emergency response teams to protect public health near contaminated sites that represent serious and immediate threats. The second approach is to prevent the creation of new contaminated sites. The third policy approach is to create a program to conduct cleanups of contaminated sites and to return the remediated sites to productive uses. Some countries already have policies in place for all three of these approaches.

Even when countries have policies in place to address all three approaches, these policies have been only partially successful. Emergency response policies are usually the most successful. When officials discover a contaminated site that poses a serious health threat, emergency actions can be direct and effective.

Policies to prevent the creation of new contaminated sites are often partially successful. In countries that enact and implement these policies, they have greatly reduced the improper disposal of toxic wastes by forcing firms to be more careful in storing and handling toxic materials to prevent spills and leaks. However, despite considerable progress, spills, leaks, and improper disposal of toxic substances still occur even in countries with the most effective prevention policies. Such efforts to prevent new contaminated sites will always fail to some degree if economic incentives make it profitable for firms to allow spills, leaks, and improper disposal of toxic substances and waste products.

Developing countries have the greatest need for policies to prevent the toxic pollution that creates contaminated sites, primarily because the globalization of the world's economies and the greater use of toxic substances in relocating industries, which are shifting pollution pressures to these countries. They currently lack the administrative capabilities of the more developed countries to implement complicated command and control policy approaches to prevent toxic pollution. The developing countries may be most effective if they concentrate on developing economic incentives to control toxic pollution to supplement the command and control approach.

Policies to achieve zero discharge of toxic substances are of special importance in efforts to prevent the creation of new contaminated sites. Such approaches serve to educate and motivate the public, interest groups, and firms. However, "virtual elimination" or "sunsetting" the use of toxic substances may be more realistic than zero discharge. Most large firms have already taken the easy and immediately cost-effective steps to eliminate or reduce the use of toxic substances in their operations. At present, there are a small number of success stories in which large firms have reengineered their entire operations to eliminate toxic materials. As it becomes increasingly expensive to use and to dispose of toxic substances, more firms will invest in reengineering to minimize their use of such substances.

The final and least successful of the three policy approaches is the actual cleanup of contaminated sites. Most countries that have an environmental cleanup policy are experiencing considerable difficulty with implementation. The expense of remediating numerous sites is the greatest problem and often creates a political problem of generating sufficient funds. Even when there is cleanup, many sites remain problems because they are not put back into productive uses.

Policy approaches in these three areas—emergency protection, prevention of new contaminated sites, and cleanup of existing sites—continue to evolve. Countries with existing policies are revising them to make them more effective and more efficient. More and more, economic incentive approaches are being instituted to supplement command and control pollution prevention approaches. Countries are also developing new remediation techniques and experimenting with different ways to pay for environmental cleanup. In addition, some governments are placing greater emphasis on putting remediated sites back into productive use. Developing countries, on the other hand, must design policy that suits their unique historical and economic circumstances.

Of these three policy approaches, the promotion of environmental cleanups is the least developed and the least successful. In designing environmental policy to promote cleanup, two critical issues must be addressed: what are the objectives of environmental cleanup, and how should these costs be paid.

THE OBJECTIVES OF ENVIRONMENTAL CLEANUP

The reasons for completing cleanups of contaminated sites seem obvious, yet clarity on the objective of such policy is essential. One objective is to lower the human health risk to an "acceptable" level. An alternative objective is to remediate all sites to a condition that allows people to use the site for any potential use, for example, food production or pumping groundwater for use as public drinking water supply. Governments that establish clear objectives for their policies have passed a critical hurdle in solving their contamination and environmental cleanup problem.

Different countries have either narrowly defined or broadly defined objectives. Some narrowly establish their objective to making the sites safe to protect public health. Other countries, though fewer in number, define the objective as protecting ecosystem health. Many countries set a broader objective of making the sites safe and putting them back into productive use. These objectives determine many of the operational issues critical to these policies, such as how clean is clean and how to evaluate whether the policy's achievements justify its cost.

The United States has a narrow objective for its cleanup policy of protecting human health and the environment. The Superfund program spends large sums on each site in attempting to make it safe for any future use. Some of the outcomes of this approach are fascinating. Despite huge cleanup costs, almost none of the remediated Superfund sites is in productive use of any kind. Most former Superfund sites are now vacant lots, often with fences surrounding them to keep people off. It is surprising that a country would choose to spend so much money on sites and then find no use for them following cleanup. The narrow objectives of the Superfund program cause this strange result. Although it addresses health concerns, it ignores the economic and social problems associated with contaminated sites.

The United Kingdom's approach to environmental cleanup has the broad objective of making sites safe and productive. After spending large sums of money for cleanups, the objective is to ensure that the country benefits from subsequent productive uses of the sites. This broader policy objective is common in the Western European countries, but requires an expanded range of work beyond the actual remediation of contamination. This expanded scope may include the following: identifying new uses for the site; identifying potential private developers; providing infrastructure or financial incentives to make the project attractive to developers; or forming public–private partnerships to develop the site. In the United Kingdom, the environmental cleanup team must identify a developer and a use for the site as a precondition for the cleanup. In The Netherlands, the reuse potential is an important factor in prioritizing remediation projects.

Broadening the objectives of environmental cleanup, however, has financial implications, for there are significant costs associated with remediation that has productive use as its ultimate goal. Even if remediation costs are held constant, the extra economic development costs can substantially increase the overall project cost. This additional cost will either increase the cost of the policy or reduce the number of contaminated sites that the program can remediate.

CONTROLLING THE COST OF ENVIRONMENTAL CLEANUP

Most of the countries with broad objectives for their cleanup policies place controls on cost. Reasons for governments to control the total expense of environmental cleanup are both internal and external to the country.

One internal reason is that a well-controlled program may help make it a priority among all of the country's other environmental efforts, especially when there are budget restrictions on environmental expenditures. In theory, programs designed to solve more important environmental problems should receive more funds than those designed to solve less important problems.

One approach to prioritize environmental protection programs is to evaluate them in terms of their ability to achieve the greatest protection, or greatest risk reduction. Governments should allocate funds that are commensurate with the relative importance of the problem and the effectiveness of the program. There is some evidence that contaminated sites receive more attention than environmental problems that pose greater risk to human health.

Countries also have external reasons to control costs, such as international trade. High cleanup costs can negatively influence a country's trade competitiveness. This is especially true for certain industries that in some countries may bear an especially high burden of funding environmental cleanup. These industries will be at a competitive disadvantage in selling their products when competing with firms based in countries with no cleanup program or a program with a different funding mechanism that does not tax private firms.

Cost control offers other advantages as well. It can increase the number of site cleanups or clean up the same number of sites but also return the sites to productive use. Countries can perform cleanup and enable subsequent use for less than the cost of remediating contamination to the maximum extent feasible, by simply allowing different levels of contamination to remain on the site after completion of cleanup. In this case, the site will not be sufficiently clean to support *all potential uses* of the site, but will be safe for the intended use.

A related issue is how to spend the money once the allocation has been set. If the government tells a community that they have $30 million to spend on the cleanup of a contaminated site, would it be more efficient to let the community decide how to proceed? If using a narrow objective for cleanup policy, they should spend all the money to remediate the contamination to the maximum extent possible. If they have a broad objective, they should spend it to remediate the contamination to the extent necessary to allow an identified productive use for the site after the cleanup. Depending on the policy objective, then, either approach can be successful.

Another approach to control cost is to restrict the remediation method, as some methods are much more expensive than others. The most important difference in cost is whether remediation is designed to safely contain the contaminants or to treat the contamination. Both of these approaches can protect human and ecosystem health from the actual or potential risk of exposure to toxic substances.

Remediation that contains contaminants so that they cannot migrate from the containment system is usually much less expensive than remediation to treat contaminants in an attempt to eliminate them. Containment is especially cost-effective when it is designed to contain contaminants on the site. A serious disadvantage is that containment approaches are not permanent solutions. In most countries, public opinion and environmental activists favor treatment, whereas business interests favor the less expensive containment techniques.

The use of containment seems to be growing in popularity in certain situations. In the United States, mini-Superfund programs operated by the states use containment much more frequently than the federal Superfund program. For example, New Jersey's Industrial Site Recovery Act allows cleanup to encapsulate contaminants on-site to promote reuse of sites. This is likely to increase as state programs become more interested in minimizing remediation cost as a way to redirect funds toward finding productive uses for sites.

The Netherlands has pioneered the use of criteria for containment when the ratio of treatment cost to containment cost exceeds a threshold. This technique avoids excessive treatment costs for cleanup. Countries with broad policy objectives are likely to increase their use of containment to control environmental expenditures.

Countries should also promote innovative remediation techniques that are more effective and less costly. Continued improvement in this area is one of the most promising trends in environmental cleanup. Yet countries must include incentives to use new techniques, otherwise there is a strong proclivity to use existing techniques that have predictable results and costs.

HOW TO PAY FOR ENVIRONMENTAL CLEANUP

Once a country establishes its objectives for environmental cleanup policy, the most critical question is how to pay for remediation. The two extreme options are to either have the government or the private sector pay all the cleanup costs. Most policies use a combination of these approaches. The proportion that the private sector pays for cleanup compared to that paid by the government is a critical question in developing efficient and equitable policy.

An approach in which the public sector pays the cost of cleanup has the powerful argument of efficiency in its favor. Public sector funding is efficient because government tax systems can efficiently collect the needed funds. The alternative, raising cleanup funds from the private sector, requires the considerable expense of acquiring the money from private firms. These money acquisition costs, including legal fees, are collectively known as "transaction costs." When private firms are forced to pay for the cleanup of some multiple-owner sites, more than half of the cost can go to transaction costs.

Yet having the private sector pay for cleanup carries two powerful arguments in its favor. The first argument is that it may be fair to hold the private firms that caused the contaminated site responsible for costs. This argument is a manifestation of the polluter pays principle. In countries where state-owned enterprises caused contamination, these enterprises are responsible for the cleanup. The polluter pays principle applies to both state-owned and privately owned firms. Since the general public believes this is fair, policies based on this principle have a great deal of political support.

The second argument in favor of private sector responsibility is functional. By forcing responsible private firms to pay for cleanup, governments create a powerful deterrence to creating new contaminated sites through intentional actions or through unintentional but careless activities. Such prevention is an essential part of any effort to solve the contamination problem.

Although most countries have laws intended to prevent the creation of sites contaminated with toxic substances, many are not able to effectively enforce these laws. Even in countries that most efficiently implement pollution prevention policies, firms will often create new contaminated sites. Policies using the polluter pays principle have the benefit of a powerful economic incentive for responsible practices, and this approach is especially important for developing countries with limited administrative capacity.

Even though an approach that forces private firms to pay for cleanup is conceptually sound, it encounters many difficulties that can make it ineffective. Often the legal tools needed to operationalize the polluter pays principle are either inefficient or inequitable. In the period from 1988 to 1991, The Netherlands succeeded in recovering only 12% of the costs of cleanup from the private sector. The United States uses extreme legal tools in this pursuit at the cost of creating clearly unfair situations that undermine public support for the cleanup program. In the final analysis, The Netherlands, the United States, and other countries that proclaim adherence to the polluter pays principle contribute a large share of the cost of environmental cleanup from general government revenues.

PLACE-BASED POLICY

One approach to solving the contamination and cleanup problem is to use place-based policies that address interrelated issues in a specific location. A contaminated site exists within a set of location-specific problems, and sometimes these problems interact, for instance, the occurrence of contaminated sites and poor prospects for economic development. Other times such problems exist in the same location, but they are not related. An example of this would be a municipality with a contaminated site that also has a recreation problem. Though many countries use a place-based approach to contaminated sites to some extent, it may be more effective to make this the main approach to addressing environmental cleanup and other problems.

A place-based policy approach recognizes the need for local involvement and ownership of programs if they are to be successful. Local communities could bring together funding from diverse programs (general revenues, environmental cleanup, public health, housing, community development, parks, recreation, and others) to tackle their specific array of environmental, economic, and social problems. This approach decentralizes decision making and allows localities to make their own risk management decisions. It is quite possible that they might allocate funds to a different environmental problem or in a different manner than the national government. In a broad sense, place-based policy allows community-based solutions.

The degree of decentralization is important to achieving environmental cleanup. In countries with decentralized cleanup approaches, communities have a greater say in the goal and the implementation of the cleanup at each site. When communities have greater responsibility for environmental policy, they often choose a strategy that puts the site back into productive use after the cleanup. Examples of local community groups taking the initiative for cleanup and reuse of sites include the Gowanus, New York, community group that is working with a local insurance company to clean up contaminated sites. Together they have formed Remedial Capital Corporation, which will purchase contaminated land, supervise the cleanup, and underwrite insurance that will indemnify all future developers (Robbins, 1994).

Place-based policy extends beyond solving environmental problems and represents a significant aspect of the trend toward decentralization. This is the driving force behind the United States' growing use of block grants (grants combining funds to solve diverse problems) rather than categorical grants (specific program funding).

In the United Kingdom, the Derelict Land Grant Programme became the urban regeneration agency English Partnerships in 1994, which now uses a place-based approach. In that same year in the United States, President Clinton designated 72 enterprise zones and 9 empowerment zones as part of a place-based approach to solving urban problems, among them contamination and cleanup. Individual U.S. states are also experimenting with these approaches, for example, the Special Area Management Plan used to remediate contaminated land in the New Jersey Meadowlands (Ochab, 1994).

CREATING GOOD POLICY

Although unique conditions in each country may dictate some variations in environmental cleanup policy, there are some principles of policy that should be valid everywhere. The goals of cleanup policy should be to achieve a speedy, cost-effective, and fair cleanup of risk-threatening contaminated sites; to put the remediated sites back into productive use; and to discourage the creation of new contaminated sites. Each of these goals needs elaboration.

A speedy cleanup requires an efficient process. Yet the process should not ignore dangerous contaminated sites, which requires that officials have a protocol to identify contaminated sites, not just wait until someone brings a site to the government's attention. Once the government identifies a site, officials should establish the priority of cleanup based on a consistent risk assessment. If the policy uses the polluter pays principle, policy makers should clearly define liability for cleanup. This should include defined funding arrangements for "orphan" sites, hardship cases, and multiple potentially responsible party sites. It should also state the date after which liability applies. The process should strive to avoid bureaucratic delay and confusion by applying clearly stated standards and procedures.

The environmental cleanup process must be cost-effective. With the high cost of remediation and the large number of contaminated sites, countries must be efficient and cost-effective to maintain public support for cleanup programs. It can aid efficiency to establish clearly defined cleanup standards that determine what level of cleanup is necessary at sites. These standards should combine the flexibility to adjust to different circumstances with the clarity to minimize uncertainty about remediation costs. Standards may allow the option of not remediating the last 5% of contaminants at a site. This alone might save sufficient money to remove 95% of the contaminants at several other sites that might otherwise receive no cleanup. The standards should also encourage some innovative remediation and be sensitive to situations where containment may be appropriate and much less expensive than treatment. The process must minimize the amount of transaction costs relative to the total cost of the environmental cleanup. Lastly, cleanup of each site must be cost-effective in comparison to the option of spending the funds on other environmental or public health programs.

The environmental cleanup process must be fair, regarding both procedural fairness and distributional fairness. Procedural fairness means that there is fair and equal application of the laws and regulations to all parties and that the process allows adequate public participation. Fairness includes the opportunity for negotiation between polluting and enforcing parties. It also requires publicly available statements of standards set and agreements reached.

Distributional fairness means that the results are fair, for example, that the cleanup program not ignore contaminated sites in low-income or minority neighborhoods. It also requires that remediated sites in these areas not have higher levels of contamination than sites in areas inhabited by more affluent people. Application of consistent cleanup standards will help ensure that the process is fair.

The second goal of cleanup policy includes returning the site to productive use after the cleanup. Policy should strive to undo the harm caused when someone's actions created the contaminated site. Undoing that harm entails remediation of the contamination and returning the site to at least its condition prior to the contamination. Yet it does not seem logical to spend large amounts of money remediating contamination only to allow the site to lie abandoned and possibly encourage urban blight and socially unwanted activities. An integral aspect of cleanup policy should be to reclaim a site and transform it into an asset to the community.

The final part of a comprehensive cleanup policy is to prevent the creation of new contaminated sites. A country cannot solve its contamination problem until it succeeds in this area. However, this is a difficult goal to achieve. The creation, transportation, storage, use, and disposal of toxic substances and products containing them offer many possibilities for intentional or unintended leaks, spills, or improper disposal. The polluter pays principle has the great advantage of providing a strong incentive to avoid creating new contaminated sites.

Today there are no models of completely successful contaminated site and environmental cleanup policies. Countries continue to modify and improve their policies, even though each country may have some policy components that are nonoptimal. However, countries with sound policies are far ahead of the many countries that have not yet started to systematically address these problems. Around the world, it is critically important to halt the creation of new contaminated sites and to aggressively yet sensibly clean up existing contaminated sites. The problem is large and environmental cleanup will stretch well into the twenty-first century.

References

Ochab, K. 1994. Special Area Management Plan in the New Jersey meadowlands. *Environment & Development.* American Planning Association, January.

Robbins, E. 1994. Piloting innovative solutions for Brooklyn's Gowanus Canal. *Environment & Development.* American Planning Association, March.

GLOSSARY

ACTION VALUES: Generic, predetermined standards for the concentrations of chemical substances in soil or groundwater that require remedial action when they are exceeded.

ACUTE TOXICITY: Toxic effects that are intense and immediate and closely correspond to what we normally think of as instances of poisoning.

AFTERVALUE: Increase in the value of reclaimed land from its value before the environmental cleanup.

APPROACH: Guiding philosophy used to establish procedures and criteria. The approach may include broad goals, the types of information to be evaluated and their relative importance, and the methods to be used in setting criteria.

ASSESSMENT CRITERIA: Concentrations of chemical substances that are used to determine if the land (soil or groundwater) is sufficiently contaminated to require remediation.

BACKGROUND CONCENTRATION: Concentration of a chemical substance that is considered to be naturally occurring in the local environment.

BEAN COUNTER: Someone or an approach that makes decisions based only on easily quantifiable factors.

BIOAUGMENTATION: Bioremediation techniques that introduce species of microorganisms that are known to work in concert with resident microorganisms to remediate organic contaminants.

BIOREMEDIATION: Group of remediation techniques that use microbes to break down organic toxic substances.

BIOSTIMULATION: Bioremediation techniques that introduce nutrients that assist resident microorganisms to break down organic contaminants.

BLUNT POLICY INSTRUMENTS: Policies that indirectly influence potentially polluting activities, for example, establishing a tax on inputs that after some industrial activity will generate toxic wastes. Whereas a direct policy instrument would regulate the waste, the blunt instrument taxes the inputs, which only indirectly provides an incentive to reduce the volume of toxic waste. The advantage of blunt policy instruments is administrative. Instead of requiring a sophisticated regulatory agency to monitor and inspect firms, the blunt instrument can utilize the existing tax collection system.

BROWNFIELD SITES: Sites for potential redevelopment that have been previously used, but are not now in productive use. Brownfield sites are usually within urban areas. See *Derelict land* and *Greenfield sites.*

CLEANUP: Actions taken to protect humans from contamination. Remediation and cleanup are often used interchangeably, although in some countries remediation is a broader term that includes both physical actions and institutional controls, whereas cleanup refers only to controls. See *Remediation.*

CO-DISPOSAL: Allowing landfills to combine municipal solid waste with industrial hazardous wastes.

COMMAND AND CONTROL APPROACH: Environmental policy approach using direct regulation to control pollution. This approach depends on standards, permits, and licenses, as well as land use and water use controls. Monitoring and enforcement systems to ensure compliance are part of this approach.

CONTAMINANT: Any chemical substance found in the environment at concentrations exceeding a standard or background concentrations.

CONTAMINATED LAND OR SITES: Parcels of land that are contaminated with toxic substances.

CONTAMINATION: Elevated concentrations of chemical substances in the environment that are the result of human activity. Pollution often refers to an ongoing activity and contamination refers to past pollution; however, these two terms are often used interchangeably.

COST PER STATISTICAL LIFE SAVED (CPSLS): Statistic calculated to allow comparison of environmental policies by comparing how much each policy would spend on average to save an additional human life.

CRADLE-TO-GRAVE: Policy approach that uses manifests to control hazardous wastes from the point of generation to the final disposal site. These controls apply to hazardous waste generators and transporters as well as those who store, treat, or dispose of hazardous wastes.

CRITERIA: Scientific information used to establish environmental objectives.

DEEP POCKETS: Refers to potentially responsible parties who have considerable financial assets or the ability to raise revenue to pay for an environmental cleanup.

DELIST: In the U.S. Superfund program, delisting means that the government removes a contaminated site from its list of sites waiting for remediation.

DENSE NONAQUEOUS PHASE LIQUIDS (DNAPLS): Liquid pollutants that migrate downward in soil and are especially difficult to contain or remove.

DERELICT LAND: Previously developed land that has been abandoned and is no longer in productive use. Such land is also known as "brownfield sites," suggesting that these sites for potential redevelopment have been previously used, as opposed to "greenfield sites," which is land that has never been developed.

DISPOSAL CHARGES: Direct taxes or fees on hazardous wastes that may be charged at either the point of generation or the point of disposal. Disposal charges are sometimes known as waste-end-taxes.

DUE DILIGENCE: Legal concept in which a landowner has taken all reasonable steps to discover if contamination is present before purchasing land. In the United States, it can be used in an "innocent landowner defense" against environmental liability.

DUTCH LIST: List of toxic substances that specifies three levels of harmful contamination, with each level determined by the concentration of hazardous substances present.

ECONOMIC INCENTIVE APPROACH: Environmental policy approach using economic incentives to control pollution. This approach establishes economic incentives for polluters to choose their own means of pollution control.

END USES: Activities intended to take place on land after remediation of existing contamination.

ENFORCEMENT INCENTIVES: Generating economic incentives by enforcing command and control regulations. They include noncompliance fees or fines, performance bonds, and liability awards. These "costs" to firms provide the incentive to minimize future pollution.

ENVIRONMENTAL AUDIT: Systematic investigations of property for measuring, evaluating, or appraising the status or probability of contamination from previous uses of the property. In the United States, there are Phase I, II, and III environmental audits with increasing thoroughness of investigation. They are often necessary to obtain financing to purchase land or insurance.

ENVIRONMENTAL INSURANCE: Insurance against environmental liability for contamination found on land. It is usually purchased when purchasing new property.

ENVIRONMENTAL KUZNETS CURVE: Assumed relationship in which pollution rises faster than economic output at low levels of income, and then has a declining intensity at more advanced stages of economic development. Also referred to as an "inverse U-shaped curve."

ENVIRONMENTAL LIABILITY: Legal responsibility for the contamination of land, which may include responsibility for the costs of remediation and any damages caused by the contamination.

ENVIRONMENTAL SINKS: Places in ecosystems where pollutants accumulate, such as wetlands and the sediments at the bottom of rivers, estuaries, and harbors.

EXPOSURE ROUTES: Environmental pathways by which contaminants can reach and expose humans or ecosystems to risk, for example, through air, water, or the food chain.

EXTENDED POLLUTER PAYS PRINCIPLE: Concept whereby the polluter pays for pollution control and cleanup (the standard polluter pays principle), plus compensation to citizens for the damages they have suffered from the pollution. See *Polluter pays principle.*

FITNESS FOR PURPOSE: Policy approach in which environmental cleanup should be limited to a level appropriate to the present or future use of the land. See *Suitable for use.*

GOLD PLATES: Providing excessive cleanup at some sites, thereby denying resources to other contaminated sites.

GREEN TECHNOLOGIES: Remediation techniques that the public views as friendly to the environment, such as bioremediation.

GREENFIELD SITES: Land that has never been developed. Greenfield sites are usually outside urban areas. See *Brownfield sites* and *Derelict land.*

GROUNDWATER: All water contained in the void spaces in soil and rocks below the water table.

HARD END USES: Redevelopment that includes structures and economic activities, as opposed to redeveloped land that is used for parks or open space.

HAZARDOUS (WASTE OR MATERIAL): Any one of the following four characteristics make a material or waste hazardous: toxicity, corrosivity, flammability, or reactivity.

INDEMNITY: Agreement to remunerate another for loss or to protect that party against liability. An indemnity is often used in the sale of a firm to help protect the purchaser against the cost of environmental cleanup for any contaminated lands for which the purchased firm may be liable.

INFILL: Property development that takes place within an urban area rather than on its periphery. The development of land parcels that are already surrounded by developed land.

INDUSTRIAL PARK: Area of land set aside for industrial uses, also called an industrial estate. The goals include promotion of industrial development because of the availability of industrial infrastructure, economies of scale in industrial infrastructure such as waste treatment facilities, and separation of industry from residential areas.

INNOCENT LANDOWNER: Owner of contaminated land who was not involved in or responsible for the contamination.

LAND USE: Types of activities that take place on specific properties. Most countries use land use planning techniques to restrict land to certain types of uses or to prohibit certain land uses from designated areas.

LEVEL PLAYING FIELD: Situation in which conditions apply to all participants or countries equally and thus would not distort economic decisions.

LIABILITY: Legal term that means legal responsibility for the damages of an action including the cost of compensating or remediating the damage, which in this context is the cost of environmental cleanup.

LIGHT, NONAQUEOUS PHASE LIQUIDS (LNAPLS): Undissolved chemicals, often petroleum products like gasoline and other fuels that float on top of groundwater rather than dissolve (mix) in the groundwater.

LOW-TEMPERATURE THERMAL TREATMENT: Remediation technique that vaporizes volatile and semivolatile organic compounds from soil, sludges, and other solids. The heat used in this technique does not burn the contaminants, thus avoiding potentially dangerous by-products.

MIDNIGHT DUMPING: Illegal disposal of hazardous wastes in nonapproved disposal sites. Before governments implemented effective cradle-to-grave policies, some unscrupulous individuals or firms dumped loads of hazardous wastes in the dark of the night.

MINI-SUPERFUND: State-run environmental cleanup programs in the United States that are intended to clean up contaminated sites that are not on the National Priority List of the federal Superfund program.

MULTIFUNCTIONALITY: Principle stating that cleaned up land should be sufficiently free of contamination to support all potential land uses, including building diverse human structures on it, extracting groundwater from it, producing crops, and providing raw materials.

NATIONAL PRIORITIES LIST (NPL): List of the most dangerous contaminated land sites in the United States that will receive an environmental cleanup under the Superfund Program.

NEW SOURCE PERFORMANCE STANDARDS: Government standards that specify in great detail the pollution-minimizing technologies that new industrial facilities may use.

NIMBY (NOT IN MY BACKYARD): Phenomenon of the public being strongly opposed to locating certain types of facilities in their communities.

OBJECTIVE: Numerical standards or narrative statements used to guide decisions about contamination and when remediation is required.

OPPORTUNITY COST: Cost to an economy of diverting substantial resources away from other economic activities.

ORPHAN PRPS: Potentially responsible parties who can no longer be found or are insolvent.

ORPHAN SITE: Contaminated land for which no responsible party can be identified or is capable of remediating the site.

PARCEL: Defined area of land that is owned by an entity.

PERFORMANCE BONDS: Payments to regulatory authorities made before a potentially polluting activity is undertaken. If the activity meets agreed-upon standards, then the performance bond is returned.

POLLUTER PAYS PRINCIPLE: Principle stating that the party responsible for causing the pollution should bear the cost of efforts to reduce or remediate pollution. This principle is a means to distribute the costs of environmental protection or cleanup. The standard polluter pays principle requires the polluter to pay only the costs of pollution control and cleanup. See *Extended polluter pays principle.*

POLLUTION: Chemical substances released to the environment at levels exceeding standards or background concentrations and considered harmful to the natural environment and/or humans. Pollution often refers to an ongoing activity and contamination refers to past pollution; however, these two terms are often used interchangeably.

PORK BARREL: Policy or program that is enacted by the political bargaining process of spreading benefits to a large number of legislative districts.

POTENTIALLY RESPONSIBLE PARTIES (PRPS): The U.S. Superfund program specifies four classes of such parties who can be held liable for the cost of environmental cleanup of sites: (1) the current owner or operator, (2) any former owner or operator, (3) any person who arranged for disposal or treatment of hazardous substances at the site, and (4) any transporter of hazardous substances to the site.

PRODUCT CHARGE: Charge on specified products, such as oils and other lubricants, that become hazardous wastes at the end of their useful life. They provide an economic incentive to minimize the production or use of such products.

PRODUCT STANDARDS: Policy approach to control hazardous materials that places controls on the use, sale, import, export, disposal, or recovery of products.

PROPERTY: Legal concept describing the ownership of a defined area of land and the bundle of property rights that define the concept of ownership.

REDEVELOPMENT: Development of land for productive activity after remediation of the contamination that resulted from earlier developed use of the site.

REMEDIAL ACTIONS: One of two types of actions (the others are called removal actions) under the U.S. Superfund program. They are long-term environmental cleanups of contaminated land sites that are listed on the National Priorities List.

REMEDIATION: Use of techniques that reduce the risk that contamination will damage the natural environment or threaten human health. Remediation can include both institutional controls such as zoning or other land use restrictions and physical actions such as removal, containment, or destruction of contaminants. Remediation and environmental cleanup are often used interchangeably, although in some countries remediation is a broader term that includes both physical and institutional controls, whereas cleanup refers only to controls.

REMOVAL ACTIONS: One of two types of actions (the others are called remedial actions) under the U.S. Superfund program. They are short-term emergency actions intended to protect human health and the environment from serious immediate threats.

RESPONSIBLE PARTY: Party who is legally liable for conducting or paying for an environmental cleanup and in some cases any damages caused by the contamination. In some countries (for instance, the United Kingdom), the responsible party refers to a subset of liable parties who actually caused the pollution event or allowed it to occur.

RETROACTIVE LIABILITY: Legal principle used by the U.S. Superfund program that makes parties liable to pay the cost of environmental cleanup for acts that were legal at the time they occurred, but that caused contamination.

RISK ASSESSMENT: Method used to estimate the risk of exposure to contaminants that employs quantitative or semiquantitative approaches to arrive at the probability of an effect.

RISK MANAGEMENT: Process of using risk assessment techniques to evaluate the magnitude of risks and of alternative remedial actions to minimize risks.

ROLLING PROGRAMMES: Reclamation strategy under the United Kingdom's Derelict Land Grant Programme that is designed to tackle the most extensive areas of dereliction in a comprehensive manner. Local authorities receive assurance of funding over a number of years to allow a strategic approach to reclaiming derelict land.

SOFT END USE: Redevelopment that is used for parks, open space, or other uses that do not require structures.

SOIL: Unconsolidated mineral and organic material occurring above bedrock. Some definitions of soil include all liquid, mineral, and organic components of the weathered crust of the earth.

SOIL VAPOR EXTRACTION: Remediation technique that extracts contaminants from soil using pressurized air forced into the soil through wells and an extraction system under a light vacuum. This technique is used for volatile organic compounds and some pesticides.

SOURCE REMOVAL: Remediation techniques that remove the toxic contaminant from the site.

SPILLOVER EFFECTS: Economic effects beyond those intended or expected by a policy or action.

STANDARD: Legally defined numerical limit or narrative statement established to define when conditions exceed acceptable levels, for instance, when chemical substances in soil or water are considered contamination.

STEPWISE CLEANUP: Partial cleanup after which redevelopment may proceed before the final cleanup starts.

SUBSIDIARITY: Official doctrine of the European Union that member-states should make decisions at the most local level of government that is appropriate for the issue at hand.

SUBSIDIES: Government financial incentives to achieve a specific goal. They include grants, low-interest loans, and tax incentives.

SUITABLE FOR USE: British term to describe the approach that determines remediation of contaminated land based on intended future land use. Remediation, if any is required, should be designed to control any unacceptable risks to health or the environment, taking into account the actual or intended use of the site.

SUNSETTING: Policy approach for phasing out the use of persistent toxic substances to achieve virtual elimination of their use. Also known as a zero discharge policy.

SUPERFUND: The U.S. Comprehensive Environmental Response, Compensation, and Liability Act (CERCLA) of 1980 established the Hazardous Substance Response Trust Fund, which is better known as the Superfund. It is also used as the name of the U.S. policy to cause the environmental cleanup of the worst contaminated land.

SUPERLIEN: Legal rights to the priority of funds for the environmental cleanup of contaminated lands. Some states in the United States have superliens giving them priority over all other liens to ensure that the state can use whatever funds that a firm may have to pay for cleanup before other creditors get paid from whatever assets are available.

SUSTAINABLE DEVELOPMENT: Development that meets the needs of the present generation without compromising the needs of future generations.

THRESHOLD VALUE: Concentration for chemical substances below which no remedial actions are necessary. Concentrations above this value require investigation, and if they exceed an action value, they require immediate remedial action. See *Trigger value.*

TOXIC: Dangerous to human health, usually depending on the precise level of concentration, compounding, or ionization of the chemical.

TOXIC HOT SPOT: Phrase used to describe the worst contaminated land sites.

TOXIC TIME BOMB: Phrase used to describe the worst contaminated land sites.

TRANSACTION COSTS: Administrative and legal costs that are in addition to the cost of the direct remediation work in the environmental cleanup of a contaminated site.

TRIGGER VALUE: Term used in the United Kingdom to indicate the concentrations of contaminants on land. It describes two situations: (1) a threshold trigger value below which there is no significant risk of contamination and (2) action trigger value at which the contamination is judged so serious that remedial action is required. Concentrations between the two trigger values require site-specific judgment. See *Threshold value.*

VIRTUAL ELIMINATION: Policy that recognizes that zero discharge may not be technologically possible, but a goal of achieving no discharges that cause a pollution problem is realistic.

WINDFALL PROFITS: Profits that do not result from any work performed. Some believe windfall profits may accrue to people who purchase contaminated land or land near contaminated land sites when the value of the land is low because of publicity about the contamination, and then sell after the environmental cleanup is complete and land values are high.

ZERO DISCHARGE: Policy approach for phasing out the use of persistent toxic substances to achieve virtual elimination of their use. Also known as sunsetting.

INDEX

U

V

W

Z